T0302071

5G Networks

An Overview of Architecture, Design, Use Cases and Deployment

Published 2024 by River Publishers
River Publishers
Alsbjergvej 10, 9260 Gistrup, Denmark
www.riverpublishers.com

Distributed exclusively by Routledge
605 Third Avenue, New York, NY 10017, USA
4 Park Square, Milton Park, Abingdon, Oxon OX14 4RN

5G Networks / by Atahar Khan, Satya Priyo Dhar, Ramakrishnan Shanmugasundaram, Joe Chemparathy.

Routledge is an imprint of the Taylor & Francis Group, an informa business

ISBN 978-87-7004-193-5 (hardback)

ISBN 978-87-7004-206-2 (paperback)

ISBN 978-87-7004-217-8 (online)

ISBN 978-8-770-04210-9 (ebook master)

A Publication in the River Publishers Series in Rapids

5G Networks

An Overview of Architecture, Design, Use Cases and Deployment

Atahar Khan

Cisco Systems, India

Satya Priyo Dhar

Cisco Systems, India

Ramakrishnan Shanmugasundaram

Cisco Systems, India

Joe Chemparathy

Cisco Systems, India

Contents

Preface

5G is the latest advancement in wireless technology that brings significant improvements in speed, capacity, latency, and connectivity compared to previous generations. It is designed to revolutionize the way we communicate and interact with technology, enabling transformative applications and services.

5G is not merely mobile technology evolution, but the breadth is extended beyond mobile communication to address all forms of communication and its enabling new services and enhancing existing services like enhanced mobile broadband (eMBB), critical communications (CC) and ultra reliable and low latency communications (URLLC), massive Internet of Things (mIoT) and flexible network operations. 5G architecture brings a drastic change from radio network to IP transport and mobile packets core for mobile network operators (MNOs) and mobile virtual network operators (MVNOs).

5G radio has evolved from the lower to the higher frequency range, which allows a higher data rate from radio units (gNDb). IP transport backhaul distributed to fronthaul, midhaul and backhaul as the packet core is virtualized and components like data units (DU) are placed closer to fronthaul.

Considering the business values of the above evolution in mind, this book is targeted towards network architects, network engineers and consultants who are involved in designing, implementing and managing 5G networks with multiple domains involved like RAN, IP transport, private/public Telco Cloud, automation and orchestration for a communication service provider (CSP) or mobile service provider.

These major topics are the emphasis of this book:

- Comprehensive coverage that provides an end-to-end understanding of the architecture, design, and orchestration of a 5G network that can interface with 5G packet core networks, 5G RANs, IP transport networks, data centers, and Telco clouds.

- Practical examples illustrating obstacles and solutions for implementing closed-loop automation in 5G transport.
- When developing 5G Transport for some of the biggest 5G networks in the world, paying attention to factors including scale, performance, latency, security, and manageability.
- A transition strategy for 4G to 5G core and transport networks.

This comprehensive book will cover not only 5G architecture and all the underlying futuristic technologies with use cases, but will also cover securing and scaling 5G as well as emphasize the need for network programmability, orchestration and automation aspects of 5G which will help people with a diverse background in an organization.

Acknowledgements

We would like to thank our colleagues in Cisco Systems with whom we have explored and learned 5G. We thank Ankush Arora (Distinguished Architect, Cisco) for guiding us and reviewing the contents of the book. We would like to thank River Publishers for the support during the writing process.

Finally, we thank our families for their patience and co-operation during the writing of the book.

About the Authors

Atahar Khan is a Principal Architect with extensive expertise in designing, architecting, and implementing complex technologies within service provider networks. His proficiency covers mobile backhaul and its progression into xHaul for 5G networking, SR-MPLS/SRv6, and Cloud RAN. Atahar holds CCIE SP certification and has received US Patents in domains related to 5G, service provider technologies, and AI/ML. Additionally, he has authored white papers on seamless/unified MPLS, segment routing network design and migration, and network timing.

Satya Priyo Dhar is a Solution Architect for Cisco with extensive experience in service provider networks architecture, design and deployment. He holds an active CCIE, and participates in technology talks, training workshops and events. He works actively with service providers on their network transformation journey to converged SDN based programmable 5G network with segment routing, EVPN and end to end automation. He has also authored a white paper on segment routing over IPv6 based design and migration.

Ramakrishnan Shanmugasundaram is a Senior Solution Architect with extensive experience in designing, architecting, and implementing complex technologies within service provider networks. His proficiency covers SDWAN, 5G networking, and SR-MPLS/SRv6. Ramakrishnan has received US Patents in domains related to IOT and service provider technologies. Additionally, he has authored white papers on segment routing networks and SDWAN use cases.

Joe Chemparathy is a Senior Solution Architect in the packet core domain with experience ranging from 2G to 5G. He has been involved in core deployments across the globe including multiple Tier-1 operators. His expertise lies in proposing suitable solutions for SP customers for both brownfield and greenfield deployments. His focus areas are Telco cloud, Cloud Native and E2E 5G architecture. He has also been involved in private 4G/5G deployments for enterprise use cases. Joe holds a master's degree in Mobile Communications from Telecom-Paris, France.

CHAPTER

1

5G Introduction

5G, short for fifth generation, is the latest advancement in wireless technology that brings significant improvements in speed, capacity, latency, and connectivity compared to previous generations. It is designed to revolutionize the way we communicate and interact with technology, enabling transformative applications and services.

5G is a new paradigm delivering up to 20 gigabits-per-second (Gbps) peak data rates. Mobile user demands for more bandwidth has gradually increased over the past few years. New business use cases are not limited to mobile users, but also extend to machines and bringing new demands like low latency and mass scale communication.

5G is not merely mobile technology evolution, but the breadth is extended beyond mobile communication to address all forms of communication and its enabling new services and enhancing existing services like enhanced mobile broadband (eMBB), critical communications (CC) and ultra reliable and low latency communications (URLLC), massive Internet of Things (mIoT) and flexible network operations. 5G architecture brings a drastic change from radio network to IP transport and mobile packets core for mobile network operators (MNOs) and mobile virtual network operators (MVNOs).

5G radio has evolved from the lower frequency to higher frequency range, which allows a higher data rate from the radio unit (gNDb). IP transport backhaul distributed to fronthaul, midhaul and backhaul as a packet core is virtualized and components like data units (DU) are placed closer to fronthaul.

3GPP has made a major contribution to defining 5G deployment standards and it is not only defining the air interface but also all the protocols and network interfaces that enable the entire mobile system: call and session control, mobility management, service provisioning, etc. The 3GPP approach helps to develop an inter-vendor and inter-operator context in 5G the networks.

1.1 Salient Features of 5G

5G wireless technology introduces several transformative features that enable advanced communication and connectivity. With significantly higher data rates, 5G offers ultra-fast download and upload speeds, revolutionizing how we consume and share content. The ultra-low latency of 5G ensures near real-time communication, making applications like autonomous vehicles and remote surgeries a reality. Another standout feature is the ability to support massive device connectivity, allowing for the seamless connection of a vast number of devices, from smartphones and IoT devices to smart cities and industrial automation. Furthermore, 5G provides enhanced network reliability and energy efficiency, making it a game-changer for mission-critical applications and sustainable infrastructure. Overall, 5G wireless technology offers unparalleled speed, responsiveness, and scalability, paving the way for innovative applications and services across various industries. 5G wireless technology brings several salient features that set it apart from previous generations. Here are some of the key features of 5G (Figure 1.1):

Figure 1.1: Features of 5G.

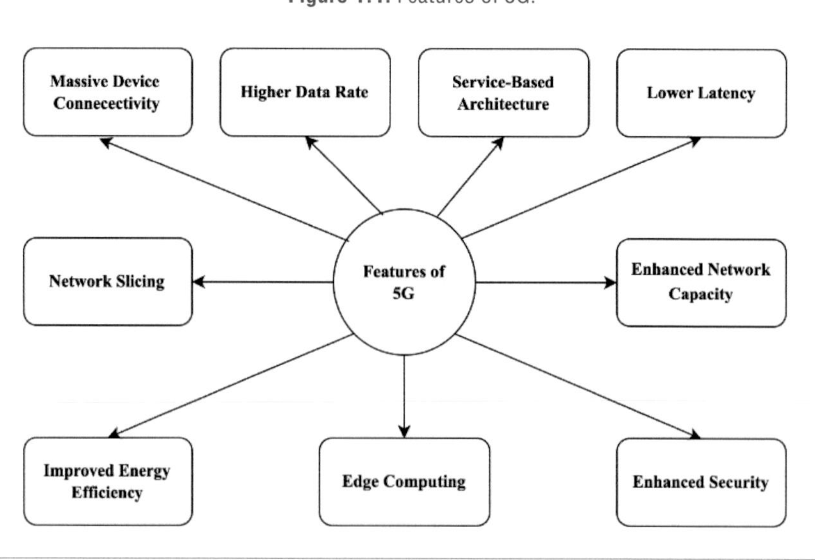

Higher data rates: 5G offers significantly higher data rates compared to its predecessors. It can provide peak data rates of up to 10 Gbps, enabling faster downloads, streaming, and real-time communication.

Lower latency: 5G aims to deliver ultra-low latency, reducing the time it takes for data to travel between devices and the network. It offers latency as low as 1 millisecond (ms), which is crucial for real-time applications like autonomous vehicles, remote surgery, and augmented/virtual reality.

Massive device connectivity: 5G is designed to support a massive number of connected devices. It offers higher device density and capacity, allowing for seamless connectivity and efficient management of Internet of Things (IoT) devices and smart city infrastructure.

Enhanced network capacity: 5G addresses the increasing demand for network capacity by utilizing advanced technologies like millimeter-wave (mmWave) frequencies, massive MIMO, and beamforming. These technologies enable higher network capacity to accommodate more devices and data traffic.

Network slicing: 5G introduces network slicing, which allows the network infrastructure to be partitioned into multiple virtual networks. Each network slice can be customized to meet the specific requirements of different applications, services, or industries, ensuring optimal performance, security, and resource allocation.

Improved energy efficiency: 5G incorporates energy-saving mechanisms, such as sleep modes and optimized signaling procedures, to improve overall network energy efficiency. This is crucial for supporting the proliferation of IoT devices and reducing the environmental impact of wireless networks.

Edge computing: 5G architecture leverages edge computing capabilities, bringing computing resources closer to the network edge. This enables faster processing, reduced latency, and improved reliability for applications that require real-time data analysis and low-latency responsiveness.

Enhanced security: 5G focuses on enhancing network security with features like stronger encryption, improved authentication methods, and network slicing isolation. It aims to provide a more secure and reliable communication environment, especially as the number of connected devices and potential vulnerabilities increase.

Service-based architecture: 5G adopts a service-based architecture (SBA), which enables more modular, flexible, and scalable deployment of network functions. It promotes the use of standardized interfaces and protocols, facilitating easier integration of new services and applications.

These features collectively enable 5G to support a wide range of transformative applications, including autonomous vehicles, smart cities, remote healthcare, industrial automation, immersive media experiences, and more. It represents a significant leap forward in terms of speed, capacity, responsiveness, and connectivity, enabling new possibilities in various industries and improving overall user experience.

1.2 5G Architecture Overview

The end-to-end architecture of 5G encompasses the entire network infrastructure, from the user equipment (UE) to the core network. It is designed to provide a flexible, scalable, and efficient framework for delivering the advanced capabilities of 5G. Figure 1.2 gives an overview of the 5G end-to-end architecture:

Figure 1.2: 5G architecture overview.

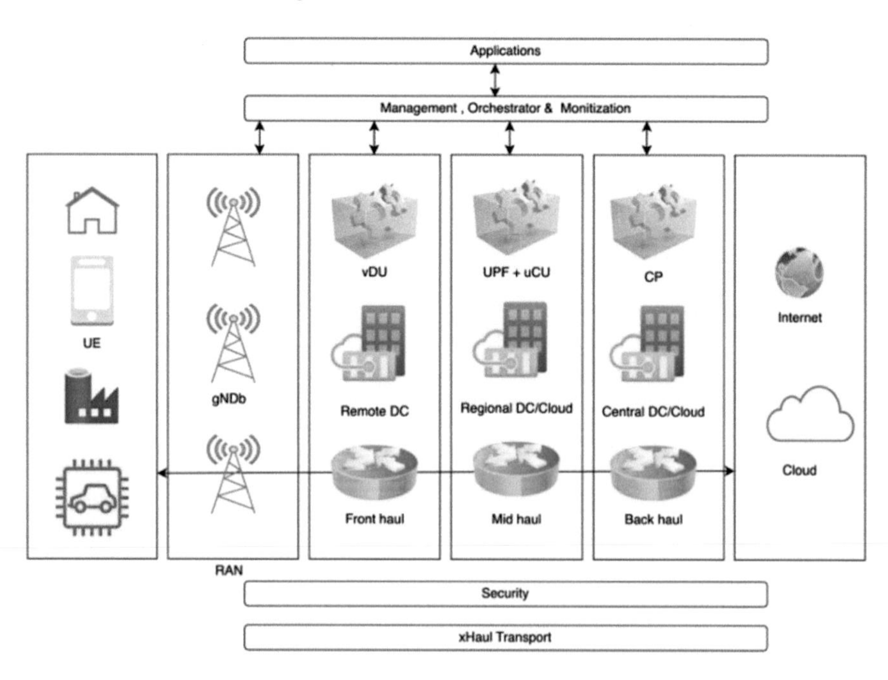

RAN networks connect to various types of access like UE, IOT cars/drones, connected homes/device/factories, etc. with high frequency channels. Packet core applications are distributed into vDU (virtual data unit), vCU (virtual control unit), UPF (user plane function) and CP (control function). vDUs are placed closer to users/UE to provide low latency at high data rate. Xhaul transport provides connectivity between the end points with traffic engineering capability. Security and automation is given utmost priority as an end to end focus area.

User equipment (UE): The UE represents the end-user devices such as smartphones, tablets, IoT devices, and other wireless devices that connect to the 5G network. These devices communicate with the network infrastructure to transmit and receive data.

Radio access network (RAN): The RAN is responsible for establishing a wireless connection between the UE and the core network. It consists of base stations, antennas, and other equipment deployed in different geographical areas. The RAN utilizes advanced technologies like beamforming, massive MIMO, and millimeter-wave frequencies to enhance coverage, capacity, and data rates.

Packet core network (PCN): The core network is the central part of the 5G architecture that handles various functions, including authentication, billing, mobility management, and routing. It is designed with a service-based architecture (SBA) and leverages virtualization technologies like network function virtualization (NFV) and software-defined networking (SDN) to provide flexibility, scalability, and efficient network management.

Network Slicing: One of the key features of 5G architecture is network slicing. Network slicing allows for the creation of multiple virtual networks within a single physical network infrastructure. Each network slice can be customized to meet the specific needs of different applications or industries. Network slices can be customized with dedicated resources, quality of service parameters, and security measures to meet the requirements of different services, such as enhanced mobile broadband (eMBB), ultra-reliable low-latency communication (URLLC), and massive machine-type communication (mMTC). For example, a network slice could be dedicated to providing low-latency connections for autonomous vehicles, while another slice could be optimized for high-bandwidth applications like virtual reality. Network slicing will enable 5G networks to support a wide range of use cases and applications, all with different performance requirements.

Edge Computing: Another important aspect of 5G architecture is the use of edge computing. Edge computing involves placing computing resources closer to end users, reducing latency and improving overall network performance. By bringing processing power closer to the edge of the network, applications can be executed more quickly and efficiently. This is particularly important for applications that require real-time data processing, such as autonomous vehicles or remote healthcare monitoring. Edge computing will be a fundamental component of 5G architecture, enabling the development of innovative and latency-sensitive applications.

Service-based architecture (SBA): The 5G architecture adopts a service-based approach that breaks down network functions into modular and interoperable components. This allows for flexible deployment, management, and integration of network functions, enabling the rapid introduction of new services and applications.

Virtualization: Virtualization technologies, such as network function virtualization (NFV) and software-defined networking (SDN), play a vital role in 5G architecture. They enable the virtualization of network functions, allowing them to run on general-purpose hardware and be dynamically deployed, scaled, and managed. Virtualization improves resource utilization, scalability, and cost-efficiency in the network. 5G workloads can either be placed in a private Telco cloud or in public cloud as a hybrid model.

The 5G architecture is expected to be based on a distributed network model, with a focus on virtualization and cloud computing. This means that network functions will be virtualized (Figure 1.3) and moved to the cloud, allowing

Figure 1.3: 5G virtual infrastructure.

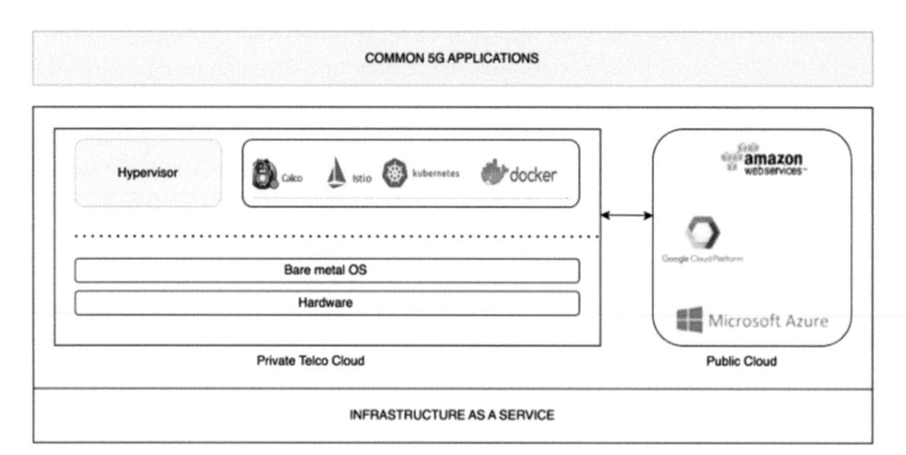

for greater flexibility and scalability. The architecture will also rely on small cell deployments, which will provide coverage in densely populated areas and improve overall network capacity. These small cells will be connected by a dense network of fiber optic cables, enabling high-speed connections and low latency.

Network functions: Various network functions are deployed in the 5G architecture to ensure the smooth operation of the network. These functions include baseband processing, radio resource management, mobility management, session management, authentication and security functions, policy control, and charging functions, among others.

These building blocks work together to create a comprehensive 5G architecture that supports higher data rates, lower latency, increased capacity, and a wide range of services and applications.

1.3 5G Building Blocks

1.3.1 5G RAN

5G RAN, also known as the next generation radio access network, is a crucial component of the 5G network infrastructure. RAN refers to the wireless network that connects end-user devices to the core network. With the rapid advancements in technology and the increasing demand for high-speed connectivity, 5G RAN has emerged as a game-changer in the telecommunications industry. It is designed to provide significantly faster data rates, lower latency, and increased capacity compared to its predecessors. The next generation 5G RAN is expected to revolutionize various industries, including healthcare, transportation, manufacturing, and entertainment.

Another significant advantage of 5G RAN is its ultra-low latency. Latency refers to the time it takes for data to travel from the source to the destination. In applications that require real-time responsiveness, such as autonomous vehicles and remote surgery, low latency is crucial. The next generation 5G RAN aims to achieve latency as low as one millisecond, enabling near-instantaneous communication between devices. This will open up new possibilities for innovative applications and services that were previously not feasible with older network technologies.

The radio access network (RAN) is a critical component of the 5G architecture that provides wireless connectivity between user equipment

(UE) and the core network. It is responsible for transmitting and receiving radio signals, managing radio resources, and ensuring reliable and efficient communication. Here are some key aspects of the 5G RAN:

gNodeB (gNB): In 5G, the base station is called the gNodeB (gNB). It interfaces with UEs and serves as the access point for wireless communication. The gNB supports advanced features such as massive MIMO (multiple input multiple output) and beamforming to enhance coverage, capacity, and data rates.

New radio (NR): The 5G RAN is based on the new radio (NR) air interface, which introduces several improvements over previous generations. It supports higher frequency bands, including mmWave, for increased capacity and data rates. NR also incorporates flexible numerology, enabling the customization of subcarrier spacing and time slot configurations to accommodate diverse use cases.

Moreover, 5G RAN offers significantly higher data rates compared to its predecessors. With speeds reaching up to 20 Gbps, users can expect lightning-fast downloads and uploads, seamless streaming of high-quality videos, and a superior overall browsing experience. The next generation 5G RAN achieves these high data rates through advanced modulation schemes and wider bandwidths. This increased capacity is particularly beneficial for bandwidth-intensive applications such as virtual reality, augmented reality, and 4K video streaming.

Massive MIMO: 5G RAN utilizes massive MIMO technology, which involves deploying a large number of antennas at the gNB. This enables beamforming and spatial multiplexing techniques to improve signal quality, coverage, and capacity. One of the key features of 5G RAN is its ability to support a massive number of connected devices simultaneously. As the number of Internet of Things (IoT) devices increases exponentially, the current network infrastructure struggles to handle the massive influx of data. The next generation 5G RAN is built to address this challenge by utilizing advanced technologies such as massive MIMO (multiple-input multiple-output) and beamforming. These technologies enable efficient use of spectrum resources and improve network capacity, allowing for seamless connectivity, even in densely populated areas.

eCPRI (enhanced common public radio interface): As the requirements of 5G networks evolved, traditional CPRI started to face limitations in terms of bandwidth, latency, and flexibility. This led to the development of eCPRI, which builds upon the foundation of CPRI but introduces enhancements to address these limitations. Key features of eCPRI include:

Split architecture: eCPRI introduces a split architecture where the baseband processing is divided into two parts: the radio processing unit (RPU) and the distributed unit (DU). This split allows for more efficient processing and better support for multi-vendor deployments.

Functional split options: eCPRI defines multiple functional split options, such as eCPRI Option 7.x, which specifies the distribution of baseband processing functions between the RPU and DU. These options provide flexibility and scalability to optimize the deployment based on the specific network requirements.

Lower latency and bandwidth optimization: eCPRI aims to reduce the latency and bandwidth requirements compared to CPRI. This is achieved by optimizing the protocol and the amount of data exchanged between the RPU and DU, enabling faster and more efficient communication.

Support for fronthaul networks: eCPRI is designed to support different transport technologies for fronthaul connectivity, including Ethernet-based solutions. This allows for greater flexibility in network deployment and management.

The adoption of eCPRI in 5G networks enables more efficient and flexible radio baseband processing, reducing latency, increasing bandwidth, and supporting multi-vendor interoperability. It plays a crucial role in the evolution of mobile networks and the deployment of advanced 5G services and applications.

Network slicing: RAN supports network slicing, allowing the partitioning of resources to create virtual networks tailored to specific applications or services. This enables the allocation of dedicated resources and quality of service guarantees for different use cases, such as enhanced mobile broadband, ultra-reliable low-latency communications, and massive machine-type communications.

Dual connectivity: 5G RAN supports dual connectivity, allowing UEs to connect simultaneously to multiple gNBs or to a combination of 4G and 5G networks. This enables seamless handover and improves coverage, capacity, and overall user experience.

Cloud RAN: Cloud RAN (C-RAN) is an architectural approach in which the baseband processing functions of multiple gNBs are centralized in a cloud-based central unit (CU). C-RAN improves resource utilization, scalability, and flexibility in deploying and managing the RAN infrastructure.

These features and capabilities of the 5G RAN contribute to the enhanced performance, capacity, and flexibility of 5G networks, enabling the delivery of high-speed data, low-latency communications, and support for a wide range of use cases and applications.

For high bandwidth availability and a latency of 1 ms, which are main objectives of 5G networks, one of the most important factors to address is the current and future RAN architecture, the link that the radio sites use to connect to the transport network as current network is not efficient enough to cater these needs. The second challenge is the manageability of this network as it is geographically dispersed.

Traditionally, the RAN has been implemented as a distributed RAN (D-RAN) architecture, where all the radio functionality (NodeB or eNodeB) resided in the cell site and exchanged IP packets with the mobile core via the xhaul network. The IP protocol used to carry the traffic streams and the data rates required were roughly equivalent to the combined user data rates of user equipment that was using the radios hosted in that cell site. An alternative design solution to the D-RAN is the disaggregated or centralized RAN (C-RAN), in which the upper layers of the radio functions are separated from the lower layers and moved to a shared, centralized location.

As shown in Figure 1.4, the 5G RAN architecture, RU (radio unit), DU (distributed unit) and CU (centralized unit) are the three main layers of

Figure 1.4: 5G RAN architecture.

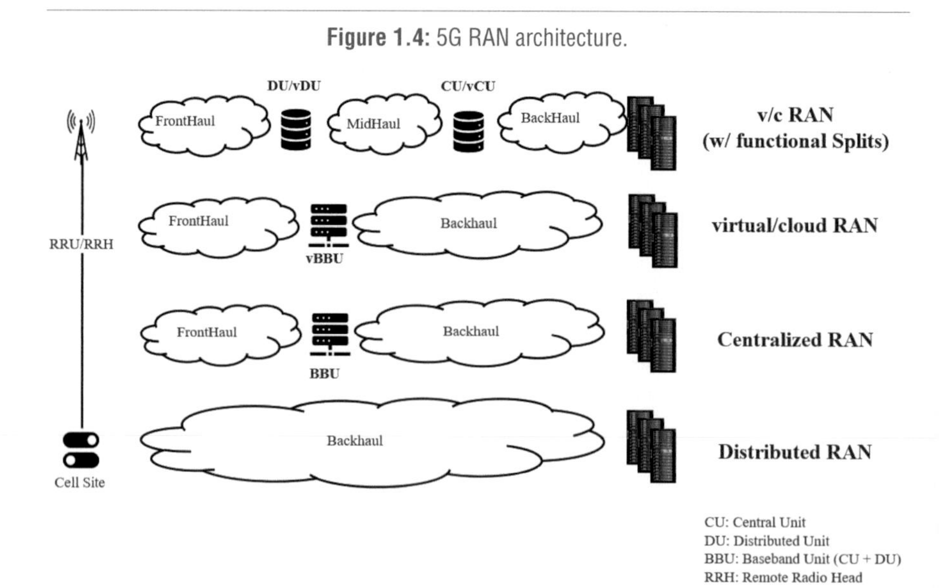

mobility communication which connects mobility partner with the service provider network wirelessly for different types of services. It carries all the 3G, 4G, 5G traffic streams from one end to the other and connected the user for the desired services.

Figure 1.4 5G RAN architecture depicts the evolution and possible design options available within the RAN network architecture. The first part of the network clearly shows that distribution of the RU, DU, CU layers together at each cell site connecting mobile gateways in the D-RAN network. Further the D-RAN network connects with backhaul which is a pre-aggregation network to the IP core.

The next section of the diagram defines the split between RU and DU, CU, which is called disaggregated or centralized RAN or C-RAN. The DU and CU layers are placed at a central location anywhere from the cell site itself out to some distance away (typically up to 10–20 km). The network that connects these RUs at cell sites to DU and CU networks is called fronthaul. The mobile fronthaul network is the connectivity between remote radio head (RRH) at the edge of a cellular network and its aggregation point, which is the centralized radio controller, also known as the base band unit (BBU). It complements the backhaul connections between the core network and the base band unit. This network is a traditional TDM based circuit switched network.

The third section shows another way of using C-RAN which includes a split between the RU, DU and CU. The network that connects the DU and CU is called midhaul, which is an Ethernet based solution; however, it does not carry the speed required for 5G.

The fourth part of the diagram shows the next generation evolved C-RAN network which splits the three sections such as the RU, DU and CU connecting via the fronthaul, midhaul and backhaul networks. It's a radio over Ethernet/Fibernet based network which carries IP traffic. This IP packet-based network can make it possible to provide high speed and high bandwidth with 1 ms latency. Of course, it needs further optimization with programmability and traffic steering for better control and capacity planning.

1.3.2 5G IP transport

5G IP transport is a critical component of the 5G network architecture that provides the foundation for carrying data traffic and supporting the advanced capabilities of 5G. IP transport enables the efficient transport of packets across the network, facilitating high-speed, low-latency, and reliable connectivity for a wide range of services and applications. With the evolution of 5G, IP transport networks have undergone advancements to meet the demanding requirements of the network. This includes the adoption of technologies such as Ethernet, MPLS (multiprotocol label switching), and IP routing protocols. These technologies enable the transport of data packets with enhanced speed, efficiency, and flexibility.

One of the key aspects of 5G IP transport is the convergence of fronthaul and backhaul transport. Fronthaul refers to the connection between base stations and the core network, while backhaul refers to the connection between base stations and the rest of the network. By converging both types of transport onto IP-based networks, operators can simplify network architecture and improve scalability, flexibility, and cost-effectiveness.

Furthermore, 5G IP transport networks support network slicing, allowing the creation of virtual networks customized to specific services or applications. Each network slice can have its own distinct requirements for bandwidth, latency, and quality of service, enabling operators to provide differentiated services and optimize network resources accordingly. Quality of service (QoS) mechanisms and traffic engineering techniques are also implemented in 5G IP transport to ensure that different types of traffic receive the appropriate prioritization and performance levels. This is crucial for delivering a seamless user experience and supporting a variety of use cases, including real-time applications and high-bandwidth services.

Moreover, 5G IP transport networks are closely integrated with edge computing and cloud infrastructure. This enables the placement of compute and storage resources closer to the edge of the network, reducing latency and enabling efficient processing of data and services. It also facilitates the deployment of edge applications and supports low-latency use cases, such as autonomous vehicles, augmented reality, and IoT applications.

One solution that has gained significant attention is segment routing. By incorporating segment routing into the 5G IP transport architecture, network operators can effectively meet the evolving needs of 5G networks. Segment routing is an innovative approach to forwarding packets in IP networks. It allows packets to be steered along a predetermined path by predefining the sequence of segments or nodes to be traversed. This flexibility and control make segment routing an ideal choice for the complex and dynamic nature of 5G networks. With the ability to define specific paths and policies, segment routing enables operators to optimize network resources, enhance performance, and deliver a more tailored user experience.

Segment routing (SR) is a network routing technology that can be utilized in the context of 5G networks to provide efficient and scalable routing solutions. It offers several benefits for 5G deployments, including network simplification, traffic engineering, and network slicing support. Here's an overview of how segment routing can be applied in the context of 5G:

Network simplification: Segment routing allows for simplified network architectures by leveraging the concept of source routing. Instead of relying on complex routing protocols and maintaining large forwarding tables at every router, segment routing encapsulates the routing instructions within the packets themselves. This eliminates the need for a per-flow state and reduces network complexity, making it easier to deploy and manage 5G networks.

Traffic engineering: With 5G networks carrying diverse types of traffic with varying requirements, efficient traffic engineering becomes crucial. Segment routing enables explicit routing paths to be defined using a sequence of segments encoded in packet headers. This allows operators to define and enforce specific paths for different traffic flows, enabling optimal utilization of network resources, improved quality of service (QoS), and better traffic management.

Network slicing: Network slicing is a key feature in 5G that enables the creation of multiple virtual networks within a shared infrastructure. Segment routing provides a flexible mechanism for network slicing by allowing operators to define specific segments for each network slice. These segments act as identifiers and enable the forwarding of packets along the desired slice-specific paths, ensuring isolation and dedicated resources for different services or applications.

Fast rerouting and resiliency: In 5G networks, fast rerouting and resiliency are crucial to ensure uninterrupted service delivery. Segment Routing can be combined with fast convergence mechanisms, such as Fast ReRoute (FRR), to provide rapid recovery from link or node failures. By precomputing alternative segments or backup paths, segment routing enables quick rerouting of traffic, minimizing service disruptions and enhancing network reliability.

IPv6 integration: Segment routing is natively built on IPv6 and can leverage the benefits of IPv6 addressing, such as a larger address space and simplified address assignment. With 5G networks adopting IPv6 as the primary addressing scheme, segment routing aligns well with the underlying infrastructure, making it a suitable choice for routing in 5G deployments.

Segment routing offers a flexible, scalable, and simplified routing solution for 5G networks. It enables network simplification, efficient traffic engineering, supports network slicing, and provides fast rerouting capabilities. By leveraging segment routing, operators can optimize their network resources, enhance performance, and deliver a diverse range of 5G services and applications with improved scalability and reliability.

5G IP transport is fundamental to the success of 5G networks, providing the necessary connectivity, performance, and flexibility to deliver high-speed, low-latency, and reliable communication. It plays a pivotal role in supporting a wide range of services and applications, driving innovation and digital transformation across various industries.

1.3.3 Telco Cloud and MEC

Telco Cloud and multi-access edge computing (MEC) are two important concepts that play a significant role in enabling the deployment of advanced services and applications of 5G. Here's an overview of each concept:

Telco Cloud refers to the virtualized infrastructure and cloud-based architecture that telecommunication service providers adopt to support their network functions and services. It involves the transformation of traditional, hardware-based network elements into software-defined, virtualized functions running on cloud-based platforms. Key features of Telco Cloud in the context of 5G include:

Virtualization: Telco Cloud leverages virtualization technologies to decouple network functions from dedicated hardware, enabling them to run as software instances on commodity servers. This brings benefits such as improved scalability, flexibility, and resource utilization.

Orchestration and Automation: Telco Cloud incorporates orchestration and automation mechanisms to dynamically manage and provision virtualized network functions. This allows for efficient resource allocation, rapid service deployment, and automation of network management tasks.

Network Slicing: Telco Cloud supports network slicing, which involves the creation of multiple virtual networks within a shared infrastructure. Each network slice can be tailored to specific use cases, enabling differentiated services with dedicated resources and QoS guarantees.

Service agility: With Telco Cloud, service providers can rapidly introduce and scale new services, responding to changing customer demands more efficiently. The cloud-based infrastructure facilitates the deployment of innovative services, such as IoT, augmented reality, and ultra-reliable low-latency communications.

Multi-access Edge Computing (MEC) also known as mobile edge computing, brings computing resources and services closer to the edge of the network, at the access network or base station level. It aims to reduce latency, improve performance, and enable real-time applications by processing data and running applications in proximity to the end-user. Key aspects of MEC in the 5G context include:

Edge Computing: MEC enables computation, storage, and networking resources to be deployed at the edge of the network, closer to the end-users and devices. This reduces latency by minimizing the distance data needs to travel, enabling faster response times and improved user experience.

Application hosting: MEC provides a platform for hosting and executing applications and services at the network edge. This allows for low-latency, real-time applications that require immediate processing and decision-making, such as video analytics, IoT data processing, and augmented reality.

Content delivery: MEC facilitates the caching and delivery of content at the network edge, reducing the load on the core network and improving content delivery performance. This is particularly beneficial for bandwidth-intensive applications, video streaming, and content distribution.

Proximity services: MEC enables the development of location-aware and context-aware services that leverage real-time data from the edge. For example, applications can use MEC to provide personalized recommendations based on the user's location, local conditions, or user preferences.

Network efficiency: MEC helps offload traffic from the core network by processing data and performing localized functions at the edge. This improves network efficiency, reduces backhaul traffic, and optimizes the use of network resources.

By combining Telco Cloud with MEC, service providers can leverage the benefits of cloud-based virtualization and edge computing to deliver low-latency, high-performance, and context-aware services in 5G networks. Telco Cloud provides the foundation for virtualized network functions and service agility, while MEC brings computing resources closer to the users, enabling innovative applications and enhanced user experiences.

1.3.4 Core network

The packet core network is an essential component of any mobile network as it handles the transmission of data packets between different devices and networks. With the advent of 5G, the packet core network needs to evolve to support the increased traffic and capacity requirements. The 5G core network is specifically designed to address these challenges and provide a seamless user experience.

One of the key features of the 5G core network is its ability to support a wide range of devices and applications. With the proliferation of Internet of Things (IoT) devices and the increasing demand for connected services, the 5G core network is built to handle the massive number of connections and diverse traffic types. This means that users can expect faster download and upload speeds, as well as improved reliability and performance.

In addition to supporting a larger number of devices, the 5G core network also offers enhanced security features. With the increase in cyber threats and attacks, it is crucial for networks to ensure that user data and information are protected. The 5G core network incorporates advanced security measures such as encryption and authentication protocols to safeguard user data.

The 5G core network (5GC) is a key component of the 5G network architecture that provides the control, management, and service delivery functions for 5G networks. It is designed to support the advanced capabilities of 5G, including ultra-reliable low-latency communications, massive machine-type communications, and enhanced mobile broadband. Here's an overview of the main elements and functionalities of the 5G core network:

Network functionality:

Access and mobility management function (AMF): The AMF handles access and mobility-related functions, such as device registration, authentication, session management, and mobility management.

Session management function (SMF): The SMF is responsible for session establishment, management, and termination. It handles the allocation of network resources and policies for data flows.

User plane function (UPF): The UPF handles the data plane functions, including packet routing, forwarding, and traffic inspection. It is responsible for processing and forwarding user data between the devices and external networks.

Policy control function (PCF): The PCF is responsible for policy and QoS control. It enforces policies and manages QoS parameters for different services and applications.

Authentication server function (AUSF): The AUSF handles user authentication and authorization functions, ensuring secure access to the network.

Unified data management (UDM): The UDM stores and manages subscriber data, providing functions such as authentication, authorization, and policy control.

Network exposure function (NEF): The NEF provides APIs for exposing network capabilities and services to third-party developers and applications.

Network slice selection function (NSSF): The NSSF determines the appropriate network slice for a user based on service requirements and network conditions.

Service-based architecture (SBA):

The 5GC is built on a service-based architecture, which allows network functions to interact with each other through well-defined service-based interfaces. This modular and flexible architecture enables the deployment and scaling of network functions independently, facilitating network evolution and innovation.

Network slicing:

The 5GC supports network slicing, allowing the creation of multiple virtual networks within a shared infrastructure. Each network slice can have its own specific characteristics, including different QoS requirements, resource allocation, and policies. Network slicing enables the provision of tailored services for diverse use cases, ensuring efficient resource utilization and meeting specific service requirements.

Support for edge computing:

The 5GC is designed to integrate with edge computing platforms, enabling the deployment of services and applications at the network edge. This integration reduces latency and enhances the performance of real-time applications by processing data closer to the end-user.

Support for IPv6:

The 5GC is built on IPv6 (Internet Protocol version 6), which provides a larger address space, enhanced security, and improved connectivity options compared to IPv4. IPv6 enables the seamless integration of various devices and supports the massive device connectivity requirements of 5G networks.

The 5G core network plays a vital role in enabling the advanced features and services of 5G. It provides the necessary control, management, and service delivery functions to ensure seamless connectivity, efficient resource utilization, and support for a wide range of use cases and applications.

1.3.5 Automation

The 5G automation architecture encompasses automation capabilities and frameworks within the 5G network to enable efficient management, operation, and orchestration of network resources and services. It leverages advanced technologies such as artificial intelligence (AI), machine learning (ML), and software-defined networking (SDN) to automate various processes and tasks in the network. The following points give an overview of the key components and functionalities of the 5G automation architecture:

Orchestration and management:

Network management and orchestration (NMO): NMO is responsible for managing and orchestrating the network resources, services, and functions. It automates the provisioning, configuration, and optimization of network elements and services.

Service orchestration: This involves the automation of service deployment and lifecycle management, including service creation, activation, scaling, and termination.

Resource orchestration: Resource orchestration automates the allocation and management of network resources, such as computing, storage, and bandwidth, to meet the requirements of different services and applications.

Artificial intelligence and machine learning:

AI/ML-based analytics: AI and ML techniques are used to analyze network data, monitor network performance, and detect anomalies or potential issues. This enables proactive management, predictive maintenance, and optimization of the network.

Cognitive automation: Cognitive automation utilizes AI and ML algorithms to automate decision-making processes, allowing the network to adapt dynamically to changing conditions and optimize resource allocation.

Intent-based networking: Intent-based networking (IBN) allows operators to define high-level intents or desired outcomes for the network, and the network automatically translates these intents into specific configurations and actions. It simplifies network management by abstracting the underlying complexity and automating the translation of intent into network operations.

Software-defined networking: Software-defined networking (SDN) separates the control plane from the data plane, enabling centralized control and programmability of network devices and resources. It allows for dynamic network configuration, provisioning, and optimization through software-based controllers.

Network function virtualization: Network function virtualization (NFV) virtualizes network functions, decoupling them from dedicated hardware and running them as software instances on commodity servers. NFV enables flexible and scalable deployment of network functions, accelerating service provisioning and reducing hardware costs.

The 5G automation architecture aims to streamline network operations, improve resource efficiency, and accelerate service delivery by automating manual tasks, optimizing network performance, and enabling dynamic resource allocation. It enhances the overall agility, scalability, and reliability of the 5G network, enabling service providers to efficiently manage and deliver a wide range of services and applications to end-users.

1.3.6 Security

The 5G security architecture is designed to address the unique security challenges posed by the 5G network and its associated services. It encompasses a comprehensive set of security measures, protocols, and mechanisms to ensure the confidentiality, integrity, and availability of data and communications within the 5G ecosystem. Key components and principles of the 5G security architecture are described here

Security functions:

Authentication and key management (AKM): AKM ensures the authentication and secure exchange of cryptographic keys between network entities and devices to establish secure communication channels.

Access control: Access control mechanisms define and enforce policies to regulate the access and authorization of users, devices, and applications within the network.

Data confidentiality and integrity: Encryption techniques are employed to protect the confidentiality and integrity of data transmitted over the network.

Security policy enforcement: Security policies are defined and enforced to govern the behavior of network entities and protect against unauthorized access, threats, and attacks.

Security incident management: Processes and tools are implemented to detect, respond to, and mitigate security incidents and breaches.

Security layers:

Radio access network (RAN) security: RAN security mechanisms protect the wireless communication links, including secure authentication, encryption, and integrity protection.

Core network security: Core network security ensures the security of data and signaling between various network elements, including secure tunneling, access control, and traffic segregation.

Application security: Application-level security measures are implemented to protect against application-layer attacks, such as secure API access, secure coding practices, and application-level firewalls.

Network Slicing security: Network slicing security ensures the isolation and protection of different network slices, preventing unauthorized access or interference between slices.

Identity management: Identity management mechanisms, such as subscriber identity management and device identity management, are employed to authenticate and authorize users and devices accessing the network.

Security monitoring and analytics: Security monitoring systems continuously monitor the network for potential security threats and anomalies, using analytics and machine learning techniques to detect and respond to security incidents in real-time.

Trust and assurance: Trust models and mechanisms, such as digital certificates, secure bootstrapping, and secure element protection, are used to establish trust between network entities and ensure the integrity of the network.

Compliance and standards: The 5G security architecture adheres to international standards and regulatory requirements to ensure compliance and interoperability across different networks and regions.

The 5G security architecture aims to provide robust protection against a wide range of security threats, including eavesdropping, data manipulation, identity theft, and network attacks. By implementing strong security measures and protocols at multiple layers of the network, the architecture helps safeguard the privacy and security of users and ensures the trustworthiness and reliability of 5G services.

CHAPTER

2

Demands Leading to Architecture Shift

The advent of 5G presents a significant opportunity for mobile operators to enhance their existing services and enable new business applications by either constructing, integrating, or upgrading their current infrastructure. Mobile operators are keen on maximizing the utilization of their existing infrastructure to ensure cost-effective deployment of new networks while maintaining seamless compatibility with their current services.

The primary objective of this book is to outline the practical hurdles faced by service providers during the transition from 4G to 5G. The book delves into how these challenges impact various technological aspects within the realm of 5G and proceeds to describe the necessary design adjustments in each of these domains to overcome these obstacles.

2.1 Packet Core Challenges

There are some key considerations that can result in various challenges that any service provider planning to transform from existing 4G network to 5G should plan are:

- Infrastructure planning
- Network slicing and planning for E2E slices
- MEC design
- Automation and orchestration planning
- Migration of subscribers from 4G to 5G.

However, the challenges for any transformation can be many more than the ones listed above; this book will be mainly centered around discussing in detail the above-mentioned challenges and design planning for different aspects of the network around these challenges.

2.1.1 Infrastructure planning

Some of the key planning aspects of infrastructure are

RAN and data center planning: For a true transition towards 5G and to be able to reap the benefits of 5G to attain a better ARPU (average revenue per user), one of the key aspects of optimization is the radio optimization.

The evolution of RAN aims to reduce the cost within the RAN domain by disaggregating the hardware and the software components of RAN, ultimately being able to introduce various standards-based interfaces into the RAN architecture thereby allowing the operator to be able to choose vendors separately for hardware, software and different layers of the RAN stack.

3GPP considered the split concept (DU and CU) from the beginning for 5G. In a 5G cloud RAN architecture, the BBU functionality is split into two functional units:

- Distributed unit (DU), responsible for real-time L1 and L2 scheduling functions.
- Centralized unit (CU) responsible for non-real-time, higher L2, and L3.

RAN disaggregation also leads to disaggregation in terms of data centers. What was typically a single data center architecture is now split into multiple data center based architecture where there is a center, edge and far-edge data centers hosting a variety of network functions.

The DU physical layer and software layer are hosted in an edge cloud data center or central office, and the CU physical layer and software can be co-located with the DU or hosted in a regional cloud data center.

While CUs will maintain BBU-like functionalities, DUs that are software-based will be more than RRH in terms of processing capacities.

Cluster planning: Some key considerations by a service provider while cluster planning are:

- Cluster design shall allow the smooth migration from current 4G.
- From a service availability viewpoint, a single point of failure should be avoided.
- Cluster design should be allow the deployment of combo network functions, e.g. (SMF+ PGW) and at the same time the cluster design should be able to anticipate the addition of CNFs in the future, i.e. the possibility of scale-in and scale-out on an as needed basis.
- Clusters should be easy to manage, monitor and debug.

When considering running a specific workload set, one must decide between running them on a small number of large clusters, each containing numerous workloads, or on many clusters, each with a smaller number of workloads. Below is a table outlining the advantages and disadvantages of different approaches:

When utilizing Kubernetes as the operational platform for applications, several fundamental questions arise regarding cluster usage:

- How should the cluster count be determined?
- What is the ideal cluster size?
- What should be included within each cluster? See Figure 2.1.

Figure 2.1: Pros and cons of various cluster approaches.

	Cost Efficiency	Ease of Mgmt	Resilience	Application Security
Large Shared Cluster	High	High	Micro	Micro
Cluster Per Environment	Medium	Medium	Small	Small
Cluster Per Application	Small	Small	Medium	Medium
Small Single Use Cluster	Micro	Micro	High	High

Few Clusters ↑ ↓ Many Clusters

Monitoring, logging, and tracing: Planning how to monitor, log and trace the calls within the network is a very complex mechanism in a 5G network, especially because the data centers are now distributed and multiple applications from different vendors will now be hosted on a single cluster, and, adding to this complexity, is the evolving standards and the vendor implementation, which is evolving with the standards.

It is of prime importance to the service provider to be able to monitor syslog information, network connections and activity, as well as Kubernetes-specific metadata like service accounts, container restarts and terminations, secrets sharing, security contexts, deployment configurations, service and port exposures, etc. The operator should also have the possibility of 100% traffic tapping with full monitoring visibility at the control-plane and user-plane, with related 3GPP standardized SBI & NON-SBI interfaces with required functionalities and features:

- End to end signaling correlation with unique session based identifiers/attributes.
- Session based identification based on SUPI/PEI and IMSI for U-plan and C-plan.
- Visualize signaling flow per protocol.
- Export signaling as Pcaps files with ~15–30 days retention period.
- Export CDRs as CSVs files with ~15–30 days retention period.
- Control plan and 5G SA procedures KPIs dashboards.
- User plan and subscriber level KQI dashboards.

End to end monitoring should also have enhancements with machine learning and AI to provide predictability and root cause analysis will also be of very high importance to the service provider considering the complexity of the network and the multi-vendor environment.

2.1.2 Network slicing and planning for end to end slices

Network slicing is a technology which allows the service provider to build multiple virtual networks on a shared infrastructure. Slicing provides the flexibility to the service providers to deploy their application and quickly accommodate specific requirements of diverse services such as augmented reality, online games, e-health, and others. As an emerging technology with several advantages, network slicing has raised many issues for the industry.

Network slicing needs to be implemented in an end-to-end manner to meet diverse service requirements with each slice having its own network architecture and protocols.

5G network slicing includes slicing 5G radio access network (RAN), 5G core network and even end-user devices and to realize the true essence of network slicing there has to be a slice awareness even within the transport network that connects the RAN and the core network; all these three areas have to be slice aware and serve the QoS that the slice is supposed to deliver to the end user.

3GPP allows some of the network slices to be heavy on sessions (massive IOT slice) and some slices to be delay sensitive (URLLC slice). Adding to the complication is the possibility of network resources being shared by slices in the RAN, core, or the transport domains and the nature of some slices being temporary and some being active only during a certain time of the day.

It becomes almost mandatory for the service provider that the slice orchestration, management and deletion is automated with minimal operator intervention due to the sheer complexity of the use cases and the deployment scenarios.

We will discuss further the domain specific challenges due to network slice and design considerations for the same in the coming sections of this book.

2.1.3 MEC design

MEC is key for realizing and deploying a meaningful 5G network to cater to different types of traffic and allow use cases that are latency sensitive. MEC is also key for private 5G or 5GaaS enablement.

Some of the key challenges or considerations in MEC design for any service provider are:

Discovery of edge application server: Some of the questions that need to be answered are:

- How does an MEC solution enable an App/UE to discover a suitable edge application server (EAS)?
- How does an MEC solution enable an App/UE to re-discover a new suitable EAS when the old one becomes non-optimal or unavailable to the App/UE?
- Should the App/UE need to be aware that there is an EAS in MEC?

Edge relocation: for edge relocation, planning is required as to:

- What triggers should be considered, and which functional entities trigger the changes to support service continuity.
- How and where to use and implement the MEC tools (e.g. UL-CL/BP insertion/relocation, SSC mode 2/3, AF influence on traffic routing, and LADN).

Service continuity in 4G on MEC: Considering that there will be a very good chance for the service to fall back to 4G, how can you ensure service continuity between 5G and 4G on MEC.

2.2 IP Transport Challenges

2.2.1 Infrastructure challenge

The transport network architecture is generally described in terms of metro access/aggregation/core domains. The NG-RAN of the 5G network consists of radio base stations (known as gNBs) connected to the 5G core network (5GC) and each other. The gNB incorporates three main functional modules: the centralized unit (CU), the distributed unit (DU), and the radio unit (RU), which can be deployed in multiple combinations in the metro network. The new 5G architecture supports the concepts of the terms fronthaul, midhaul, and backhaul for the transport network to support the Fx interface, F1 interface, and NG interface between 5G nodes. In addition, the Xn interface provides interconnection between different NG-RAN nodes (gNB or eNB). Fronthaul networks for standard public radio interface (CPRI) transport deployed today for 4G using dark fiber will evolve to Ethernet-based transport technology. The architecture of the 5G transport network domain with respect to IP transport is summarized as follows (Figure 2.2):

Figure 2.2: New IP transport architecture for 5G RAN.

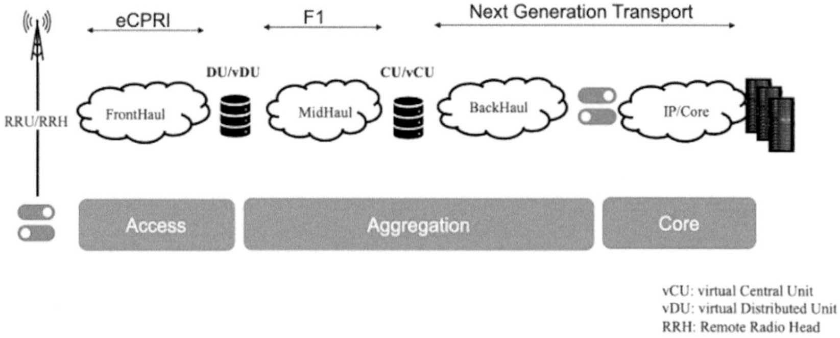

vCU: virtual Central Unit
vDU: virtual Distributed Unit
RRH: Remote Radio Head

- If DU is deployed near to RU, or RU/DU/CU are integrated. In this scenario, the distance between DU and RU is generally very short (less than 100 m) as such no transport network is required between RU and DU.
- CU deployment can be on midhaul or backhaul depending upon location of CU:
    - If CU is deployed with DU, then only a backhaul transport network is required.
    - If CU is away from DU, then both midhaul to support F1 interface and backhaul to support the NG interface are required.
- **With centralized RAN deployment**, DU is located far away from RU at a centralized location and the distance between DU and RU is longer. In this scenario, a fronthaul transport network to support the interface (CPRI/eCPRI) is required, along with either midhaul (for F1) or backhaul (for NG) or both transport networks.

- **With virtualized RAN deployment**, CU is located far away from DU at a centralized location with a data center (also called mobile edge computing (MEC)). DU may or may not be collocated with RU. If DU is not co-located with RU, then all the three fronthaul, midhaul, and backhaul will be required; else only midhaul and backhaul will be required.

Every new wireless network technology – from 3G to 4G to 5G – requires an upgrade to the IP transport network (legacy backhaul and now fronthaul/midhaul), which is used to carry network traffic from the cell site to the mobile core. However, in a 5G world, where network performance is expected to take a massive leap in capacity and bandwidth, backhaul becomes even more critical. For network transformations from 4G to 5G, capacity upgrades are inevitable as mobile network operators (MNOs) need to upgrade the metro/aggregation and core IP transport network, which may require a deployment of high-speed routers/line cards ranging from multiples of 25G to 50G in the access domain and 100G to 400G in the core domain.

2.2.2 Network clocking

Another significant change towards the 5G transformation is coming with timing and synchronization. With 5G, time division duplex (TDD) radios are being deployed alongside and among other TDD and FDD radios. TDD radios are highly dependent on a reliable phase and time synchronization source to operate correctly. These TDD radios will be driving the vast majority of the capacity at radio sites in 5G networks. Therefore, synchronization backup over backhaul is a must. Thus, the synchronization solution defined in (G.8275/Y.1369) should be used in the transport network to support 5G frequency and phase/time synchronization requirements. As per this solution, every node between the clock server and the end application node should support the SEC/eSEC and T-BC or T-TC clock (ITU-T G.8271.1). Figure 2.3 shows a clocking design for 5G transport which is a generic construct for synchronization for the transport network and depicts one example of how such a network could be designed.

Generic guidelines for the clocking design:

- In general, the frequency reference master PRC/ePRC is deployed in the core network, and the phase/time master PRTC/ePRTC is deployed in the access, aggregation, or core network.
- For the frequency synchronization solution, the transport nodes between the PRC/ePRC and RRU shall support the appropriate SEC or eSEC physical layer clock.
- For the phase/time synchronization solution, the transport nodes between the PRTC/ePRTC and RRU shall support the T-BC PTP layer clock. The clock specification is (ITU-T G.8273.2), the

Figure 2.3: Clocking design for 5G transport.

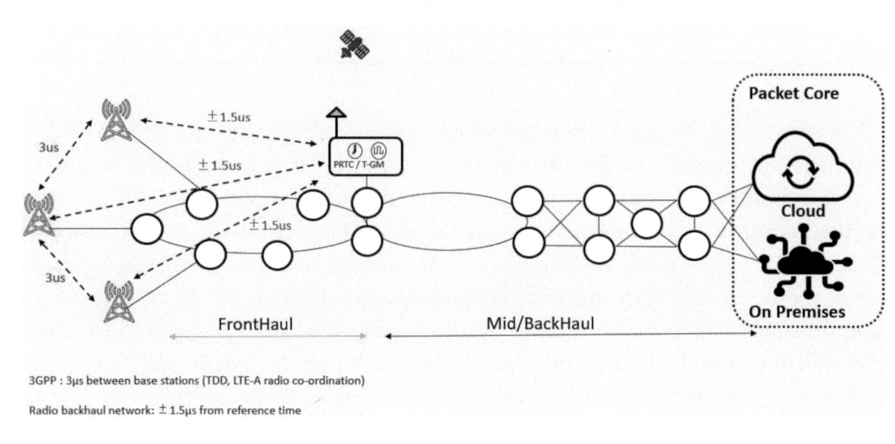

network limit is defined in (ITU-TG.8271.1), and the PTP full timing support profile is (ITU-T G.8275.1).

- Optical layer nodes without optical protection/restoration are not required to support the OEC, eOEC or T-BC. This is because these nodes do not affect the accuracy of transport synchronization network.

2.2.3 Converged SDN transport

To scale and optimize the network infrastructure cost, for 5G readiness, we recommend mobile network operators (MNOs) design their network with converged SDN transport design. The converged SDN transport design is built over the segment routing solution and offers simplicity, scalability and reduce complexity for large-scale MNOs. Furthermore, it leverages the principles of seamless unified MPLS by decoupling the transport and service layers of the network, thereby allowing these two distinct entities to be provisioned and managed independently and seamlessly interconnecting the access, aggregation, and core MPLS domains of the network infrastructure with hierarchical segment routing label-switched paths (LSPs).

Converged SDN transport design introduces an SDN-ready architecture which evolves traditional metro network design towards an SDN enabled, programmable network capable of delivering all services (residential, business, 4G/5G mobile backhaul, video, IoT) on the premise of simplicity, full programmability, and cloud integration, with guaranteed service level agreements (SLAs).

Figure 2.4: 5G End to end converged SDN transport.

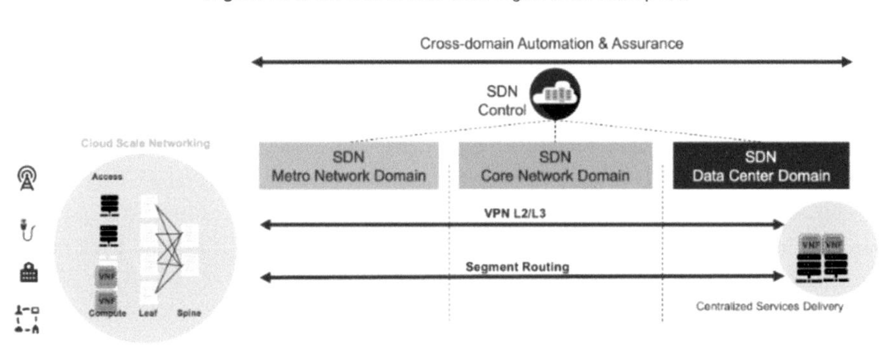

The converged SDN transport design (Figure 2.4) satisfies the following criteria for scalable next-generation networks:

Simple: Based on segment routing as a unified forwarding plane, and EVPN and L3VPN as a standard BGP based services control plane.

Programmable: Using SR-PCE to program end-to-end multi-domain paths across the network with guaranteed SLAs.

Automated: Service provisioning is fully automated using open models; analytics with model-driven telemetry in conjunction to enhance operations and network visibility.

2.2.4 Convergence

As compared to 4G, 5G is expected to carry more traffic, and some of them are very critical, for example, autonomous vehicles and remote surgery. Therefore, it calls for better protection and restoration of its transport network. Protection or restoration mechanisms should be used in the 5G transport network as necessary to meet the requirements of the services being carried over the 5G network.

Mobile network operators (MNOs) need to implement the technologies to accomplish rapid convergence from the transport level standpoint. It focuses on fast recovery from failure using the transport layer.

2.2.5 MEC challenges for IP transport

The 4G LTE hardware networks were based on a hub-and-spoke design, susceptible to network congestion, and not aligned with ultra-low latency requirements. With SDN-based 5G networks, the notion is to move away from hub and spoke IP transport physical architecture towards the distributed architecture whose design intentionally eludes the congestion bottlenecks. For delay-sensitive use cases, MEC deployment is a critical technology that helps to reduce latency by deploying MEC applications on top of the MEC server at the edge of the network, as discussed at the start of the document. From a transport network perspective, to reduce the latency, MNOs can deploy the distributed MEC deployment contrary to hub and spoke design. By following this approach, it will alleviate the latency requirements from the transport network perspective.

2.2.6 IP transport slice for 5G services

Network slicing is a method of creating multiple isolated logical and virtualized networks over a common multi-domain physical infrastructure, tailored to fulfil diverse requirements requested by a particular application. Network slicing uses network virtualization to divide single network connections into multiple distinct virtual connections that provide different amounts of resources as per the need for different types of traffic, i.e. supporting strict SLA requirement as well as best effort traffic over the same network infrastructure; hence guaranteeing service assurance to end customer.

Network slicing is fundamentally an end-to-end partitioning of the network resources and network functions so that the selected applications/end user services may run in isolation from each other to effectively meet the desired business outcome. A network slice can be dynamically created, modified or deleted without impacting other slices.

In order to achieve slicing at scale in a transport network, a catalogue of predefined slice forwarding templates will need to be built and instantiating the most suitable template when a slice request is placed. With this approach, potentially hundreds or thousands of "slice instances" can be automatically instantiated using a limited set of "slice types". A transport network slicing architecture is comprised of three fundamental components:

- A collection of control points in the user plane that have the ability to steer selected traffic flows on one path vs. another.
- Data enablement of the control point so a policy engine can control the traffic flow path selection.
- A management and orchestration system that creates the slices, configures them for operation, and manages them.

2.3 5G Security

5G offers the potential for the next stage in mobile connectivity, characterized by enhanced reliability and low latency. However, this advancement also introduces a host of new vulnerabilities and potential attack vectors that service providers must carefully address in their deployment strategies.

2.3.1 RAN Security

5G radio access network (RAN) security is a pivotal element in ensuring the integrity, confidentiality, and reliability of wireless communications. As the interface between user devices and the core network, the RAN is a critical point where security measures must be robust.

Encryption plays a central role in RAN security. 5G networks mandate the use of advanced encryption protocols, safeguarding data as it travels between user devices and base stations. This encryption ensures that sensitive information remains confidential, shielding it from potential eavesdropping or interception.

Authentication is another cornerstone of RAN security. It involves verifying the identity of both user devices and base stations before granting access to the network. By employing robust authentication mechanisms, the RAN prevents unauthorized devices from connecting, safeguarding the network against potential threats.

Furthermore, RAN security extends to the protection of network equipment and infrastructure from physical and cyber threats. Implementing access controls, monitoring for anomalies, and regular security updates are essential practices to maintain a secure and resilient RAN infrastructure.

Overall, robust RAN security measures are crucial to the successful deployment and operation of 5G networks, ensuring that users can enjoy the benefits of high-speed, low-latency connectivity without compromising on data security and privacy.

2.3.2 IP Transport Security

IP transport security is a foundational element in the architecture of fifth-generation wireless networks. As 5G relies extensively on IP-based communication, ensuring the confidentiality, integrity, and availability of data in transit is paramount.

One of the key aspects of IP transport security in 5G is the implementation of strong encryption mechanisms. Data traveling over IP networks is encrypted to safeguard it from potential eavesdropping or tampering. This encryption ensures that sensitive information, including user data and critical network functions, remains protected throughout its journey.

Access control and authentication are also crucial components of 5G IP transport security. Network devices, including routers and switches, must be protected against unauthorized access. Secure authentication protocols verify the identity of these devices and ensure that only trusted entities can interact with the network.

Moreover, IP transport security encompasses measures to mitigate distributed denial-of-service (DDoS) attacks and other network-level threats. These measures include traffic filtering, rate limiting, and intrusion detection systems to detect and respond to anomalies in network traffic.

5G IP transport security is fundamental to the reliable and secure operation of 5G networks. It encompasses encryption, access control, authentication, and measures against network-level threats, all working together to provide a robust security framework for the data flowing through 5G networks.

2.3.3 Data center security

5G data center security is a critical consideration in the architecture of fifth-generation wireless networks, especially with the increasing reliance on cloud-native and virtualized technologies. These data centers form the backbone of 5G networks, housing the infrastructure responsible for processing and managing the massive volumes of data and services.

One of the primary aspects of 5G data center security is access control. Strict access policies and authentication mechanisms are employed to ensure that only authorized personnel can access data center resources. This helps prevent unauthorized tampering or breaches of sensitive data.

Virtualization security is also paramount in 5G data centers. The use of virtual machines (VMs), containers, and microservices requires robust isolation mechanisms and continuous monitoring to prevent any compromise within the virtualized environment.

Additionally, data center security in 5G networks involves advanced threat detection and mitigation mechanisms. Intrusion detection systems (IDS), intrusion prevention systems (IPS), and security information and event management (SIEM) solutions are used to identify and respond to potential threats in real time.

Physical security is not to be overlooked, as 5G data centers house critical hardware components. Access controls, surveillance, and environmental controls are essential to protect the physical infrastructure from unauthorized access, natural disasters, and other physical threats.

5G data center security is a multi-faceted approach that encompasses access control, virtualization security, threat detection, and physical security measures. It plays a pivotal role in ensuring the reliability, availability, and integrity of the data and services that underpin 5G networks, contributing to a safe and robust telecommunications ecosystem.

2.3.4 Packet core security

Packet core security is a foundational element in ensuring the integrity, confidentiality, and availability of data and services within 5G wireless networks. The packet core, often referred to as the heart of the 5G network, serves as a central hub for routing and processing data traffic, making its security paramount.

One of the key aspects of 5G packet core security is the implementation of robust encryption protocols. These protocols are applied to the data packets as they traverse the core network, ensuring that sensitive information remains confidential and protected from potential eavesdropping or tampering.

Access control and authentication mechanisms are also vital components of packet core security. User devices, network functions, and services must be authenticated before gaining access to the core network. This prevents unauthorized entities from entering the network and helps maintain its integrity.

Moreover, 5G packet core security involves proactive measures to detect and mitigate threats. Intrusion detection systems (IDS) and intrusion prevention systems (IPS) continuously monitor network traffic for anomalies and known attack patterns, enabling rapid response to potential security breaches.

5G incorporates new networking concepts and adapts existing ones to its architecture. Some of these architectural transformations involve modernizing mobile architecture to align with cloud operations, while others stem from emerging use cases like augmented and virtual reality, MEC, CRAN, and network slicing. The 5G core architecture introduces fresh security challenges due to the virtualization of mobile network components and the creation of distinct slices for various 5G use cases. It also exposes mobile network core elements to third-party applications and external Internet-facing interfaces using 3GPP-specified exposure functions like service capability exposure function (SCEF) and network exposure function (NEF). Another aspect of the security landscape is the external-facing interfaces, such as peering points and roaming interfaces, which facilitate interconnections among different operators, enabling their subscribers to roam seamlessly.

With these new architectural shifts and connection points, the threat surface in 5G has expanded.

In conclusion, robust 5G packet core security measures are essential to the safe and reliable operation of fifth-generation wireless networks. Encryption, access control, authentication, and threat detection mechanisms work in tandem to create a secure and resilient core network, ensuring that users can enjoy the full benefits of 5G connectivity without compromising on data security and privacy.

2.4 Service Orchestration and Assurance

Service orchestration and assurance in 5G networks bring a set of unique challenges due to the complexity and dynamic nature of these networks. Here are some key challenges in this domain:

2.4.1 Network Slicing management

5G networks support network slicing, which allows the creation of isolated, customized network segments for various services and applications. Managing and orchestrating these slices efficiently while ensuring they meet the service-level agreements (SLAs) and quality of service (QoS) requirements is a significant challenge. Service orchestration platforms need to dynamically allocate and optimize network resources for each slice. Implement end-to-end network slicing orchestration. The network slicing solution must be cross-domain by design to stitch together an end-to-end slice spanning the RAN, edge, transport, and the core.

2.4.2 End-to-end service assurance

Ensuring end-to-end service quality and performance across a multi-vendor, multi-technology network is complex. Service assurance tools and processes must seamlessly integrate with various network elements, protocols, and domains to provide holistic visibility and control. Detecting and troubleshooting issues that span different parts of the network is a major challenge.

Service lifecycle management: Managing services through their entire lifecycle, from creation and provisioning to monitoring and optimization, requires a cohesive approach. Orchestration and assurance solutions must handle service onboarding, scaling, updating, and decommissioning efficiently, considering dependencies and interrelationships.

Dynamic resource allocation: 5G networks are characterized by their ability to allocate resources dynamically based on real-time demand. Service orchestration platforms need to be agile in resource allocation and de-allocation to optimize network efficiency and meet SLAs while avoiding resource contention.

Security and privacy: Ensuring the security and privacy of network services is a paramount concern. Service orchestration and assurance systems must incorporate robust security mechanisms to protect against cyber threats, unauthorized access, and data breaches.

Cross-domain orchestration: 5G networks often span multiple domains, including radio access, core network, cloud infrastructure, and edge computing. Coordinating and orchestrating services seamlessly across these domains while maintaining performance and SLAs is a significant challenge.

Interoperability: Achieving interoperability between different vendors' equipment and software is a common challenge in 5G networks. Service orchestration and assurance systems should support open standards and interfaces to ensure compatibility and avoid vendor lock-in.

Scaling and elasticity: 5G networks need to accommodate varying levels of demand efficiently. Orchestration and assurance systems must be capable of auto-scaling services and resources to handle fluctuations in traffic while maintaining service quality.

Regulatory and compliance: Staying compliant with regional and international regulations, such as data privacy laws and spectrum management policies, is crucial. Service orchestration and assurance systems should include features for monitoring and enforcing compliance.

Addressing these challenges in 5G service orchestration and assurance requires a combination of advanced technologies, industry collaboration, and continuous adaptation to the evolving network landscape. Solutions that can provide end-to-end visibility, automation, and proactive issue detection will be essential for ensuring the reliability and performance of 5G services.

2.4.3 Disaggregated systems (HW from SW)

Disaggregated systems in 5G networks present a dual-edged challenge for service orchestration and assurance. On one hand, they offer network operators unprecedented flexibility, innovation opportunities, and resource efficiency by allowing the mixing and matching of components from different vendors. This flexibility can lead to more dynamic and cost-effective network management. However, the very attributes that make disaggregated systems appealing also introduce complexities and uncertainties that must be effectively managed.

The challenge lies in orchestrating and assuring services across a heterogeneous environment where various vendors' components need to work together seamlessly. Ensuring interoperability, end-to-end service quality, and security in such a diverse ecosystem can be daunting. Network operators must grapple with complexities ranging from resource allocation and troubleshooting to vendor management and compliance. To navigate this challenge successfully, service orchestration and assurance solutions must evolve to accommodate the intricacies of disaggregated systems. These solutions should provide comprehensive visibility, automation, and real-time monitoring capabilities to maintain service quality and reliability in a disaggregated 5G network landscape. Additionally, industry collaboration and the development of standardized interfaces will be pivotal in addressing the unique challenges posed by disaggregation while harnessing its potential benefits for the next generation of telecommunications.

2.4.4 Decomposed systems (virtualization, containers, microservices)

The adoption of decomposed systems, including virtualization, containers, and microservices, in 5G networks presents a host of intricate challenges for service orchestration and assurance. These architectural paradigms break down network functions into smaller, more agile components, offering benefits like enhanced flexibility and scalability. However, they also introduce a layer of complexity in managing and optimizing services.

One of the primary challenges is orchestrating services across a diverse landscape of virtualized network functions, containers, and microservices. These components can span multiple domains and vendors, each with its own interfaces and protocols. Ensuring seamless integration and interoperability between these building blocks while delivering end-to-end services that meet stringent quality of service (QoS) and service-level agreements (SLAs) demands a sophisticated orchestration framework.

Security remains a paramount concern. Decomposed systems require robust security measures to protect against cyber threats and unauthorized access to individual components. Each virtualized network function, container, or microservice represents a potential entry point for attackers, necessitating comprehensive security policies, monitoring, and threat detection mechanisms.

In conclusion, while decomposed systems offer the promise of greater network agility and resource efficiency in 5G, they come with a set of intricate challenges for service orchestration and assurance. Successfully navigating these challenges demands advanced orchestration platforms that can handle the complexities of managing diverse components, adapting to dynamic changes, maintaining security, and ensuring compliance while delivering uninterrupted, high-quality services in the rapidly evolving 5G landscape.

2.5 DC Transformation

2.5.1 DC transport integration challenges

One of the major challenges today with respect to transport between the data center and service provider is scale and automation. Traditionally the server provider has a scaled setup and extending multiple VPN/VRFs between data center and transport MPLS routers requires either IP hand-off achieved via VRF-lite or VXLAN encapsulation. However, the key challenge involved with this solution is that each of the VPN networks needs to be extended manually, and with multiple VRFs, the complexity increases proportionally with more sub-interfaces (one for each VRF). In addition, we would require configuring and maintaining multiple instances of routing adjacencies, which is not scalable as the network grows. For using VXLAN encapsulation, the service provider transport device needs to support the VXLAN which is not typical in service provider environments.

The other important requirement of 5G is network slicing and automation. As customers transform their network to 5G, it is important to have a functionality of automated network slicing such that the policies can be created for each 5G services and are propagated from DC to the transport network seamlessly.

2.5.2 DC fabric with SR-MPLS

The data center SDN solution is a key aspect that helps address these challenges on the journey towards 5G transformation. Using DC fabric with SR-MPLS features enables unified transport between border leaf switches to the SP core routers seamlessly without major changes to the encapsulation protocols used in the existing environment. This feature helps customer scale without many changes in the core infrastructure as they can continue using the SR/MPLS in the core. With SR/MPLS handoff from DC, we can now form a single BGP EVPN session between the DC border leaf switch and the handoff device to exchange all the prefixes in all the VRFs. This provides end to end segmentation of DC VRF EVPN routes which is translated to MPLS VPNv4 at the handoff. This makes the overall process of deployment, operations and scale seamless and simple. The feature enables integration with automation and orchestration framework easier. We can also achieve consistent policy construct such as the prefixes can be advertised with specific color communities or with DSCP/EXP marking and the provider edge router can apply SR policy in transport using these color community or DSCP/EXP markings. The SR/MPLS handoff also helps achieve automation in network slicing, such that each slice could be part of same VPN or different VPN and that the traffic within a slice could use one or more class of service in transport based on the requirement. Additionally, the incoming traffic can also be marked by border leaf based on EXP values thereby prioritizing the traffic within the fabric as required.

2.5.3 Flexible service chaining

In today's Telco world, a typical mobile service provider has various service functions for data traffic offering security and optimization services. This can range from TCP optimizers, parental control to deep packet inspections, and CGNAT appliances as part of Gi-LAN service components. These services are mainly deployed as independent service functions whether PNF or VNF, from a wide range of vendors and makes service chaining static and bound to the network topology for insertion and policy selection. This leads to operational and management complexity as part of day 2 operations. This traditional approach towards service insertion comes at a cost in terms of complex troubleshooting, maintenance and meet scale on demand. Further with 5G transformation, where the services are placed at the edge locations closer to the user, this approach further gets complicated with decentralized management and complex automation/orchestration process.

The two most important challenges are:

Service chaining deployments are tightly coupled: This limits service agility, especially in a virtual environment. Implementing service nodes into an application path, independent of location, has been a challenge for many service providers. Think about scenarios where you have to make traffic flow through service nodes (for example CGNATs) or multiple services nodes (example CGNAT and TCP Optimizer), you usually build that manually and involving complex PBRs on both network and service devices to steer the traffic through multiple service nodes. Requirements to make changes on the fly, make certain traffic bypass a particular node or a service and scale-in and scale-out further complicates the design.

Lack end-to-end service visibility and cross-domain orchestration complexity: A network operator must invest in monitoring and management of multiple service nodes from various vendors individually from network perspective resulting in increased operational expenses. Additionally, the automation and cross-domain orchestration pose complexity in terms of integrating with each service node individually. The complexity increases many-fold when looking at from distributed architecture where the services are hosted at different locations in Telco DC architecture.

2.5.4 Symmetric traffic for Gi-LAN service

DC PBR can load-balance traffic to multiple PBR nodes for a specific service and traffic can be redirected to one of the PBR nodes based on hashing. Because most of the service devices perform connection tracking, they must see both directions of a flow. Therefore, you need to make sure that incoming and return traffic are redirected to the same PBR node. Symmetric PBR is the feature that enables this capability, The use case for symmetric PBR with the hash tuple is a scenario in which the traffic from a source IP address (user) always needs to go through the same service node.

CHAPTER

3

5G Use Cases

5G technology opens a wide range of use cases and applications across various industries due to its high-speed, low-latency, and massive device connectivity capabilities. Here are some key 5G use cases:

3.1 Enhanced Mobile Broadband (eMBB)

5G eMBB (enhanced mobile broadband) is one of the primary use cases for 5G technology (Figure 3.1). It focuses on delivering significantly faster data speeds, higher capacity, and improved network performance for mobile broadband services.

5G eMBB (enhanced mobile broadband) represents a transformative use case within the realm of 5G technology. At its core, eMBB is about delivering unprecedented data speeds and network performance to mobile broadband users. With the capability to offer multi-gigabit download speeds, 5G eMBB is set to revolutionize how we experience connectivity on our mobile devices. This means faster downloads, seamless streaming of 4K and 8K video content, and lightning-fast web browsing. Beyond speed, eMBB is characterized by low latency, ensuring that real-time applications like online gaming and video conferencing are incredibly responsive and immersive. Moreover, 5G eMBB networks boast high capacity and can accommodate a vast number of devices, making it ideal for crowd events and IoT deployments. Whether it's enabling augmented and virtual reality experiences, powering fixed wireless access in underserved areas, or enhancing remote work capabilities, 5G eMBB is poised to reshape the digital landscape and drive innovation across various industries.

Figure 3.1: 5G eMBB , mMMV and URLLC use cases.

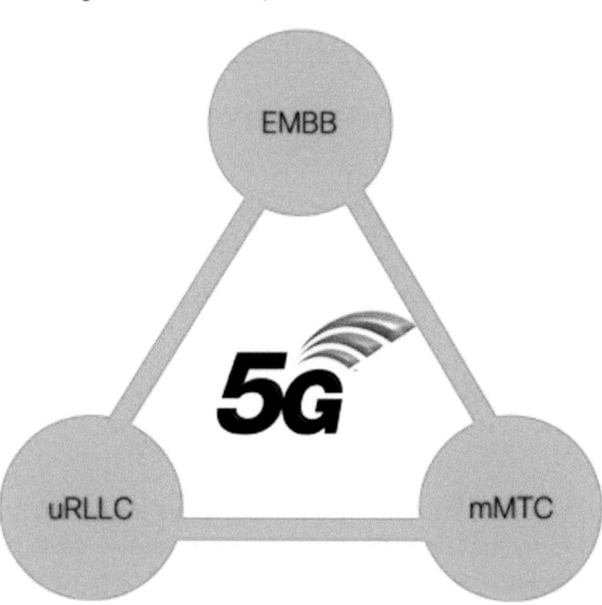

Fixed wireless access (FWA) is another example of 5G eMBB that can serve as an alternative to traditional wired broadband for residential and business users. It allows for the deployment of high-speed broadband services in areas where laying fiber or cable is challenging or expensive.

3.2 Internet of Things (IoT)

Enabling massive IoT deployments with support for a vast number of connected devices makes applications like smart cities, smart homes, and industrial IoT more efficient.

5G IoT powered by 5G technology, brings forth a myriad of transformative use cases. At its core, it excels in connecting and managing a vast array of devices with unmatched efficiency. This capability finds applications across numerous domains. In industrial settings, 5G IoT facilitates the real-time monitoring and control of machines, leading to enhanced productivity and predictive maintenance. In smart cities, it underpins intelligent traffic management, environmental monitoring, and public safety enhancements, making urban living more sustainable and secure. Precision agriculture leverages 5G IoT to optimize crop yields through soil and weather data analysis. In healthcare, remote patient monitoring becomes seamless, while in logistics, supply chain management benefits from real-time tracking of goods and resources. The automotive sector is revolutionized by connected and autonomous vehicles, ensuring safer and more efficient transportation. Additionally, 5G IoT contributes to improved energy grids, personalized retail experiences, environmental conservation, and much more. With its high-speed, low-latency connectivity, and massive device support, 5G IoT is ushering in a new era of connectivity and innovation across industries, ultimately reshaping the way we live and work.

These use cases demonstrate the versatility of 5G IoT in transforming industries and improving various aspects of our daily lives. The combination of high-speed connectivity, low latency, and massive device support makes 5G IoT a foundational technology for the future of the Internet of Things.

3.3 Augmented Reality (AR) and Virtual Reality (VR)

5G technology is poised to revolutionize the realm of augmented reality (AR) and virtual reality (VR), offering immersive experiences and innovative use cases. The unparalleled speed and low latency of 5G networks are pivotal for the seamless functioning of AR and VR applications. In the realm of AR, users can expect real-time, context-aware overlays of digital information onto the physical world. This opens the door to applications such as smart glasses for navigation, industrial maintenance, and education. 5G-powered VR, on the other hand, delivers stunningly immersive environments and experiences, from gaming and entertainment to training simulations and virtual meetings. It also enables collaborative VR experiences where users from different locations can interact seamlessly, making it a powerful tool for remote work and education. These technologies are not limited to entertainment but extend to healthcare, where surgeons can perform remote surgeries with precision, and education, where students can explore historical sites or complex scientific concepts firsthand. With 5G AR and VR, the boundaries between the physical and digital worlds blur, offering limitless possibilities for innovation and transformative experiences.

In the realm of augmented reality (AR), 5G's high-speed, low-latency connectivity enables real-time, context-aware overlays of digital information onto the physical world. This has applications in navigation, where AR can provide dynamic, step-by-step directions overlaid on a user's view as they move through unfamiliar environments. In industrial settings, AR can facilitate maintenance and repair tasks by superimposing instructions, diagrams, and data onto machinery and equipment, enhancing efficiency and reducing errors. In education, AR can bring textbooks and learning materials to life with interactive 3D models and simulations, making complex subjects more accessible and engaging for students.

Virtual reality (VR), powered by 5G, offers immersive experiences that are incredibly lifelike and responsive. This is particularly impactful in the gaming and entertainment industries, where users can dive into virtual worlds with breathtaking graphics and interact with their surroundings in real-time. Beyond entertainment, 5G-enabled VR has applications in education and training, where users can participate in realistic simulations, such as medical procedures or hazardous job training, without physical risks. In healthcare, VR can be used for therapy and pain management, providing immersive distractions and relaxation techniques to patients.

5G AR and VR also open up opportunities for remote collaboration and communication. Virtual meetings and conferences can become more interactive and engaging, enabling participants from different locations to meet virtually in a shared environment. Architects and designers can use VR to collaboratively review and modify 3D models in real-time, regardless of their physical location.

These are just a few examples of the myriad use cases for 5G AR and VR. The combination of high-speed, low-latency connectivity and immersive technology promises to transform industries and revolutionize the way we work, learn, entertain, and communicate in the near future.

3.4 Autonomous Vehicles

Enabling real-time communication among autonomous vehicles, traffic management systems, and infrastructure for safer and more efficient transportation.

The integration of 5G technology with autonomous vehicles represents a transformative use case with profound implications for the future of transportation. With its ultra-low latency and high-speed data transmission capabilities, 5G serves as the backbone for enhancing the communication and safety aspects of autonomous vehicles. These self-driving cars rely on real-time data exchange with their surroundings, including other vehicles, traffic infrastructure, and pedestrians. 5G's V2X (vehicle-to-everything) communication enables this seamless interaction, allowing vehicles to share critical information and make split-second decisions to prevent accidents. High-definition mapping, precise localization, and sensor fusion further augment the vehicle's ability to navigate complex environments with precision. Additionally, 5G enables remote monitoring, fleet management, and over-the-air updates, ensuring that autonomous vehicles are always up to date and safe. In essence, 5G technology unlocks the full potential of autonomous vehicles, making them safer, more efficient, and capable of transforming the way we move and commute in the future.

5G technology is pivotal in realizing the potential of autonomous vehicles by providing the high-speed, low-latency communication infrastructure required for safe and efficient self-driving operations. This use case highlights the immense impact of 5G on the future of transportation, where autonomous vehicles are set to revolutionize mobility, reduce accidents, and increase overall efficiency on the roads.

3.5 Smart City

The integration of 5G technology into smart cities represents a monumental shift in urban development, creating a dynamic, connected, and sustainable urban environment. Here's a detailed exploration of some key use cases for 5G in smart cities (Figure 3.2):

Figure 3.2: 5G Smart city use cases.

Smart traffic management: One of the stand-out use cases for 5G in smart cities is the revolutionization of traffic management. 5G's low latency and high bandwidth enable real-time data collection from sensors, cameras, and vehicles. This data is used to optimize traffic signal timings dynamically, reducing traffic congestion, and minimizing commute times. Moreover, 5G powers vehicle-to-everything (V2X) communication, enabling vehicles to communicate with each other and with traffic infrastructure. This not only enhances road safety but also paves the way for the widespread adoption of autonomous vehicles, which can navigate cities efficiently and safely.

Environmental monitoring: 5G-enabled environmental monitoring is another critical aspect of smart cities. Air quality sensors connected through 5G continuously measure pollution levels, providing real-time data for city authorities to take swift action to improve air quality. Additionally, weather stations benefit from 5G's high-speed connectivity, ensuring rapid and accurate transmission of weather data. This empowers cities to better prepare for extreme weather events, mitigate risks, and bolster disaster management strategies.

Energy efficiency: Smart grids are a hallmark of 5G-enabled smart cities. With real-time data on energy consumption and distribution, smart grids optimize energy usage, reduce wastage, and seamlessly integrate renewable energy sources. Residents and businesses can remotely monitor and control their energy consumption, contributing to energy conservation and cost reduction. This not only benefits the environment but also enhances energy reliability and resilience in the face of outages.

Public safety and emergency services: 5G-powered surveillance cameras and sensors are strategically deployed throughout the city to ensure public safety. High-resolution cameras connected via 5G enable real-time surveillance, incident monitoring, and the swift response of law enforcement agencies. First responders benefit from 5G's high-quality video feeds and real-time data transmission, allowing them to respond faster and more effectively to emergencies, accidents, or disasters.

Waste management: Smart waste management is streamlined through 5G-enabled sensors in waste bins. These sensors signal when bins are full, optimizing waste collection routes and reducing operational costs. Additionally, interactive 5G-powered screens at recycling stations educate citizens on proper recycling practices, promoting sustainability and environmental responsibility.

Public Wi-Fi and connectivity: 5G networks provide ubiquitous, high-speed, and reliable public Wi-Fi coverage throughout the city. This ensures that residents, visitors, and businesses enjoy seamless connectivity for work, education, and leisure activities. Digital inclusion efforts can bridge the digital divide by offering affordable or free 5G-enabled internet access to underserved communities, fostering digital literacy and economic participation.

Healthcare services: 5G brings transformational changes to healthcare services within smart cities. High-quality, real-time telemedicine services become the norm, enabling residents to consult healthcare professionals remotely. Patients with chronic conditions are monitored through wearable devices connected via 5G, offering personalized, continuous care, and reducing healthcare costs.

Education: Education is greatly enhanced by 5G in smart cities. High-definition video streaming, enabled by 5G, supports remote and interactive digital classrooms, ensuring educational continuity. Augmented reality (AR) applications, powered by 5G, create immersive learning experiences, allowing students to explore subjects interactively and fostering a deeper understanding of complex topics.

Tourism and hospitality: For tourists and the hospitality industry, 5G offers remarkable advantages. Mobile apps powered by 5G provide tourists with location-based information, real-time translations, and interactive experiences that enhance their exploration of the city. Hotels and hospitality services leverage 5G to offer guests personalized services and immersive entertainment, enriching the overall tourist experience.

In essence, 5G technology empowers smart cities to redefine urban living, bolster sustainability, and elevate the quality of life for their inhabitants. These use cases represent a fraction of the transformative potential of 5G in shaping the cities of the future. By embracing 5G, cities are poised to become smarter, more connected, and more resilient in the face of the challenges and opportunities of the modern age.

3.6 Remote Surgery and Telemedicine

The integration of 5G technology into healthcare has ushered in a new era of remote surgery and telemedicine, revolutionizing the way medical care is delivered and expanding access to expertise across geographical boundaries. 5G's exceptional low-latency, high-bandwidth capabilities make it possible for surgeons and medical professionals to perform and oversee procedures from a distance with unparalleled precision and immediacy. In the context of remote surgery, 5G ensures that the slightest movement by a surgeon's hand is transmitted in real-time to a robotic surgical system, allowing for delicate and complex surgeries to be conducted remotely with reduced risk and enhanced patient outcomes.

Telemedicine, powered by 5G, enables patients to consult with healthcare providers via high-quality, real-time video and audio connections. Physicians can diagnose, monitor, and prescribe treatment remotely, improving access to care for individuals in remote areas or with mobility constraints. Moreover, the integration of augmented reality (AR) and virtual reality (VR) through 5G adds a new dimension to medical training and education, allowing medical students and professionals to engage in immersive simulations and collaborative learning experiences. With 5G, the future of healthcare is characterized by increased accessibility, reduced healthcare disparities, and the potential for more precise and efficient medical interventions, ultimately leading to improved patient care and outcomes.

3.7 Industrial Automation

The application of 5G technology in industrial automation marks a significant leap forward in the efficiency, productivity, and safety of manufacturing and industrial processes. With its ultra-low latency and high data throughput, 5G enables real-time communication and control, unlocking a multitude of use cases:

Real-time monitoring and control: 5G allows for real-time monitoring and control of industrial machines and processes. This means that factories can instantly adjust machine settings, respond to issues, and optimize production in real-time. This reduces downtime and enhances overall efficiency.

Predictive maintenance: 5G-enabled sensors continuously collect data on the condition of equipment and machinery. Advanced analytics and machine learning algorithms process this data to predict when maintenance is needed, allowing companies to perform maintenance proactively, avoid breakdowns, and extend the lifespan of their assets.

Remote expertise: 5G facilitates remote collaboration with experts. Technicians on-site can wear augmented reality (AR) glasses that provide real-time video and audio feeds to experts located elsewhere. These experts can then guide on-site personnel through complex tasks, repairs, or troubleshooting.

Robotics and autonomous machines: 5G enables the real-time communication required for the safe and efficient operation of robots and autonomous machines on factory floors. These machines can work alongside human workers, performing repetitive tasks, and handling dangerous materials, thereby improving safety and productivity.

Quality control: High-definition cameras and sensors connected via 5G allow for precise quality control and inspection of manufactured products. Defects can be detected and addressed immediately, reducing waste and ensuring product quality.

Inventory management: 5G supports the use of connected sensors to monitor inventory levels and automate restocking processes. This prevents stockouts, reduces overstocking, and optimizes supply chain operations.

Supply chain optimization: 5G-enabled tracking and tracing solutions provide real-time visibility into the movement of goods throughout the supply chain. Companies can make data driven decisions to optimize logistics and minimize delivery delays.

Energy management: Smart factories equipped with 5G can optimize energy usage by monitoring and controlling lighting, heating, and cooling systems in real-time. This leads to energy savings and reduced environmental impact.

Customized production: 5G allows for agile manufacturing processes that can quickly adapt to changing customer demands. This enables more flexible and efficient production of customized or small-batch products.

Worker safety: Wearable devices connected through 5G can monitor worker safety by tracking vital signs, detecting falls, and providing real-time alerts in case of emergencies. This enhances worker safety in industrial environments.

In summary, 5G is a game-changer for industrial automation, facilitating real-time communication, data processing, and control. It streamlines operations, improves productivity, reduces costs, and enhances worker safety. Industries are embracing 5G to stay competitive in an increasingly automated and interconnected world, paving the way for more efficient and sustainable manufacturing processes.

3.8 Retail and Advertising

The convergence of 5G technology with the retail and advertising industries is reshaping the way businesses engage with consumers. This partnership is defined by its ability to deliver rich, interactive, and highly personalized experiences. In the retail sector, 5G enables the creation of smart stores where augmented reality (AR) and virtual reality (VR) applications come to life. Shoppers can virtually try on clothing, visualize furniture in their homes, or explore products in immersive 3D, fostering a more engaging and informative shopping experience. With real-time data analytics and hyper-personalized marketing, retailers can deliver tailored promotions and recommendations to customers as they shop, creating a sense of individualized attention. Additionally, the advent of cashierless stores powered by 5G eliminates checkout lines, streamlining the purchasing process and enhancing convenience.

In the advertising arena, 5G unleashes the potential for location-based and interactive campaigns. Advertisers can precisely target consumers with promotions and content based on their real-time location, maximizing the impact of their messages. Interactive ads come to life on consumers' devices through AR and VR experiences, fostering brand engagement and leaving a lasting impression. Smart shelves and inventory tracking powered by 5G ensure that products are readily available, enhancing the customer experience. In essence, 5G is transforming the retail and advertising landscape, enabling businesses to connect with consumers in more meaningful ways, deliver seamless shopping experiences, and drive sales through innovative, data driven marketing strategies.

3.9 Smart Agriculture

The integration of 5G technology into agriculture is revolutionizing the way farmers and agribusinesses operate, fostering a new era of precision and sustainability. One of the standout use cases for 5G in agriculture is precision farming. With 5G-enabled sensors, drones, and IoT devices, farmers gain access to real-time data on soil moisture, nutrient levels, weather conditions, and crop health. This data empowers them to make highly informed decisions about when and where to plant, irrigate, fertilize, and apply pesticides, resulting in optimized resource utilization, reduced environmental impact, and increased crop yields. The ability to remotely monitor and control autonomous farming equipment via 5G ensures that tasks such as plowing, seeding, and harvesting are executed with unmatched precision and efficiency, reducing labor costs and enhancing overall farm productivity.

Another significant use case lies in livestock monitoring. Wearable devices equipped with 5G connectivity enable farmers to track the health and location of their livestock in real-time. This includes monitoring vital signs, detecting illnesses early, and ensuring the well-being of animals. This level of monitoring not only enhances animal welfare but also improves farm management by enabling proactive interventions and reducing losses. In essence, 5G technology is reshaping agriculture into a highly connected, data driven, and sustainable industry, where every aspect of the farming process benefits from real-time insights and precision, ultimately contributing to global food security and environmental stewardship.

3.10 5G Enabled Education

5G technology has the potential to revolutionize education by providing high-speed, low-latency connectivity that enhances remote learning and virtual classrooms. With 5G, students and educators can engage in high-quality, real-time video conferencing, ensuring that remote learners have access to the same level of interaction and engagement as in-person classes. This enables a seamless transition between physical and virtual learning environments, making education more flexible and accessible.

Immersive learning experiences: Virtual reality (VR) and augmented reality (AR) applications powered by 5G offer immersive learning experiences. Students can don VR headsets to explore historical sites, conduct virtual science experiments, or engage in interactive simulations. This not only makes learning more engaging but also allows students to grasp complex concepts more effectively. For example, medical students can practice surgical procedures in a virtual operating room, and history students can step into ancient civilizations through AR-enhanced textbooks. 5G's low latency ensures that these experiences are seamless and responsive, providing a truly immersive educational environment.

Instant access to educational resources: 5G enables students to access a vast array of educational resources instantly. Whether it's streaming high-definition educational videos, participating in live online classes, or accessing cloud-based educational materials, 5G ensures that students can tap into these resources without delays or buffering. This level of connectivity empowers educators to leverage a wide range of digital tools and platforms, making education more dynamic and tailored to individual learning styles. Additionally, 5G supports real-time collaboration and communication among students and teachers, fostering a sense of community and facilitating group projects, even in virtual settings.

Remote education and beyond: 5G not only enhances remote education but also extends its benefits to rural and underserved areas, bridging the digital divide. Remote schools and students in rural locations can now access high-quality educational content and participate in virtual classrooms with ease. Furthermore, 5G opens up opportunities for lifelong learning, enabling professionals to upskill and reskill through online courses, workshops, and webinars, ensuring that education remains a continuous and accessible journey throughout one's life.

5G technology is poised to revolutionize education by making remote learning more engaging, immersive, and accessible. It empowers students and educators with high-quality connectivity, enabling a seamless blend of physical and virtual classrooms and fostering a future where education knows no geographical boundaries.

3.11 Cloud Gaming and Low-latency Multiplayer Experiences

These are just some examples of the diverse 5G use cases that are transforming industries and enhancing connectivity and communication capabilities in our increasingly digital world. As 5G networks continue to expand and mature, we can expect even more innovative applications to emerge.

Cloud gaming and low-latency multiplayer experiences: The advent of 5G technology has revolutionized the gaming industry, offering gamers a host of exciting possibilities. One of the stand-out use cases for 5G in gaming is cloud gaming services (Figure 3.3). With the ultra-fast, low-latency connectivity of 5G, gamers can access and play high-end video games via cloud streaming. This eliminates the need for expensive gaming hardware and allows players to enjoy console-quality gaming experiences on a range of devices, from smartphones to smart TVs. Gamers can seamlessly stream games from remote data centers, resulting in a more accessible and cost-effective gaming ecosystem.

Low-latency multiplayer gaming: 5G's ultra-low latency is a game-changer for multiplayer experiences. Gamers can engage in fast-paced online matches with virtually no lag, leading to more responsive and enjoyable gameplay. Whether it's first-person shooters, real-time strategy games, or esports competitions,

Figure 3.3: 5G cloud gaming.

5G ensures that players' actions are reflected in the game world in real-time, enhancing the competitiveness and thrill of multiplayer gaming. This low latency also opens the door to new gaming experiences, such as augmented reality (AR) multiplayer games that seamlessly blend the virtual and physical worlds.

Cross-platform play and content creation: 5G-powered multiplayer gaming transcends platform limitations. Gamers on different devices, such as consoles, PCs, or mobile devices, can play together seamlessly. This fosters a more inclusive and diverse gaming community, where friends and family can connect and enjoy games regardless of their preferred gaming platform. Additionally, 5G facilitates live game streaming and content creation. Gamers can easily broadcast their gameplay on platforms like Twitch or YouTube with minimal latency, interacting with viewers in real time. This has given rise to a new generation of gaming influencers and content creators, further enriching the gaming ecosystem.

In conclusion, 5G technology has unleashed a wave of innovation in the gaming world, making high-quality gaming more accessible, immersive, and socially connected than ever before. Cloud gaming, low-latency multiplayer experiences, cross-platform play, and content creation are just a few examples of how 5G is shaping the future of gaming, providing gamers with a diverse range of experiences and opportunities to connect with others in the gaming community.

CHAPTER

4

RAN Architecture Design

4.1 Introduction to Today's D-RAN Architecture

Traditional D-RAN (distributed radio access network) architecture refers to a network architecture approach for radio wireless communication systems, typically used in cellular networks like 4G and 5G. In this architecture, the baseband processing units (BBUs) and the remote radio heads (RRHs) are distributed, allowing for more flexibility and scalability.

As shown in Figure 4.1, the cell site acts as a RAN boundary. The cell site connects the RAN to mobile backhaul and eventually to mobile packet core. In D-RAN, each site terminates the radio network, and each cell site is an independent entity, thus it is called distributed RAN network.

The D-RAN architecture as shown in Figure 4.1 consists of the following main components:

Centralized baseband unit (BBU): The BBU is responsible for processing the baseband signals, which include tasks like modulation, demodulation, encoding, decoding, and beamforming. In a traditional D-RAN architecture, the BBUs are centralized in a central location, often referred to as a baseband unit hotel or BBU pool. The BBUs are connected to the remote radio heads (RRHs) via high-capacity fiber optic links.

Remote radio head (RRH): The RRH is responsible for converting the baseband signals received from the BBU into radio frequency (RF) signals that can be transmitted over the air. The RRH is typically located closer to the

Figure 4.1: Distributed radio access network.

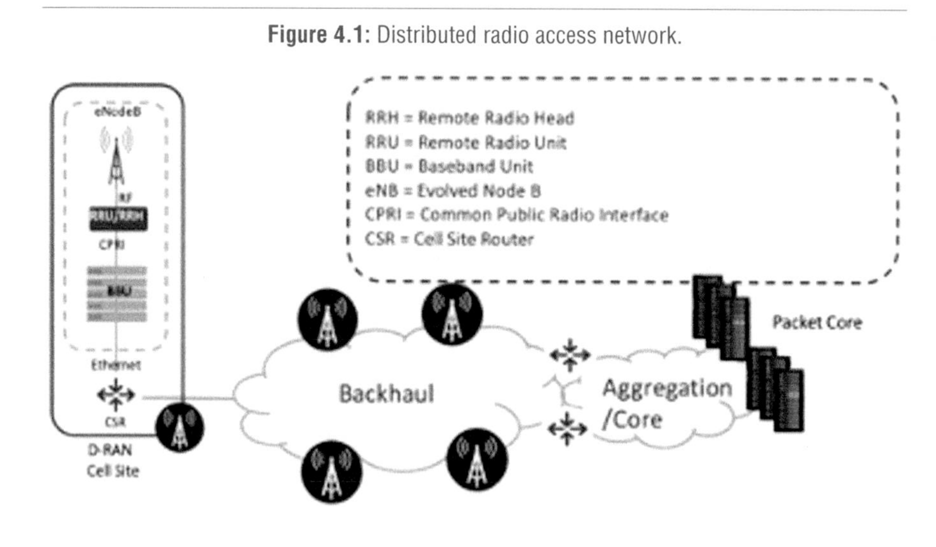

antennas, which reduces the transmission losses and improves the overall system performance. The RRHs are connected to the BBUs through fiber optic links.

Antenna system: The antenna system consists of the antennas and associated RF components, such as filters and amplifiers. The antennas are responsible for transmitting and receiving RF signals to and from the mobile devices. In a D-RAN architecture, the antennas are connected to the RRHs, which are in turn connected to the BBUs.

CPRI interface: This interface specifies the communication protocol and standards used between the BBU and RRH. A common public radio interface (CPRI) is the common fronthaul interface in D-RAN.

eNodeB: eNodeB, short for evolved node B which serves as the base station, it establishes and maintains wireless communication with user equipment (UE) such as mobile devices. The eNodeB performs various critical functions, including radio access, mobility management, resource allocation, connection management, packet routing, quality of service (QoS) management, and synchronization. It ensures efficient transmission of data, seamless handover between cells, optimal utilization of resources, and reliable connectivity for UEs within its coverage area.

Fiber optic link: This is the high-capacity link that connects the BBUs to the RRHs. It carries the baseband signals from the BBU to the RRH for RF conversion and transmission. In D-RAN, this link is typically implemented using fiber optic cables to ensure low latency and high bandwidth.

Cell site router (CSR): It serves as a gateway between the cellular base stations, such as eNodeBs or small cells, and the wider network infrastructure. The cell site router handles the routing and forwarding of data packets between the base stations and the core network. It is connected to the BBU via ethernet connection and further connects the BBU to the mobile backhaul network. It also performs functions like network address translation (NAT), quality of service (QoS) management, traffic prioritization, and network security. The cell site router plays a crucial role in enabling seamless connectivity, efficient data transmission, and reliable performance in cellular networks.

BBU pool: The BBU pool refers to a centralized location where the BBUs are housed in D-RAN architecture. It typically consists of multiple BBUs, which are shared among multiple RRHs. The BBU pool allows for efficient resource utilization, centralized management, and maintenance of the baseband processing units.

Split architecture: The split architecture defines how the baseband processing tasks are divided between the BBU and RRH. In D-RAN, there are different split options, such as the centralized split (C-split) and distributed split (D-split), which determine the amount of processing performed at the BBU and RRH respectively.

Understanding these D-RAN terminologies will help to grasp the concepts and architecture of distributed radio access networks more effectively.

The traditional D-RAN architecture offers several advantages, including:

Improved capacity and coverage: By distributing the BBUs closer to the antennas, the D-RAN architecture reduces the transmission losses and improves the overall system capacity and coverage.

Flexibility and scalability: The distributed architecture allows for easy expansion and upgrades, as new RRHs and BBUs can be added to the network as needed. This flexibility makes it easier to adapt to changing network requirements and traffic patterns.

Lower operational costs: By centralizing the baseband processing in a BBU pool, the D-RAN architecture reduces the number of BBUs required, resulting in lower power consumption and maintenance costs.

Overall, the traditional D-RAN architecture provides a more efficient and flexible solution for wireless communication systems, enabling operators to improve network performance, capacity, and coverage while reducing operational costs. D-RAN offers some of the benefits of centralization and distribution, such as improved capacity and coverage compared to a fully distributed architecture while maintaining a level of flexibility and scalability. It can be more cost-effective and easier to deploy than fully centralized RAN architectures.

4.2 Radio Frequency Primers

4.2.1 Cell sectors

In a radio access network (RAN), cell sectors refer to the division of a cell into smaller coverage areas using directional antennas. Each cell sector has its own set of antennas that transmit and receive signals in a specific direction. This allows for targeted coverage and increased capacity within the cell. The sectorization of cells is a common technique used to optimize network performance and efficiently manage resources.

Cell sectors offer several benefits in a RAN. Firstly, by dividing a cell into sectors, operators can tailor the coverage area to specific geographical locations or areas with high user density. This ensures that resources are allocated efficiently, and signal strength is optimized for users in each sector. Secondly, cell sectorization increases capacity by enabling sector-specific frequency reuse. By using directional antennas, the same frequencies can be reused in different sectors within a cell, resulting in higher network capacity and improved spectral efficiency.

Another advantage of cell sectors is the reduction of interference between neighboring cells. Directional antennas help focus the signal in a specific direction, minimizing the potential for interference with neighboring cells operating on the same frequencies. This interference reduction improves network performance, data rates, call quality, and overall user experience. Cell sectors are typically identified by sector IDs or names, allowing for easy management and configuration of the network.

In summary, cell sectors play a vital role in optimizing coverage, capacity, and performance in a RAN. By dividing cells into smaller sectors with directional antennas, operators can provide targeted coverage, increase network capacity, and reduce interference, resulting in an improved wireless communication experience for users.

Radios (carriers) per sector for each technology (2G, 3G, 4G and 5G):

- Each antenna panel in a single sector can contain two (or more) antennas.
- Each radio (carrier) from each technology (2G, 3G, 4G and 5G) is connected to at least one antenna.
- Low band radios (carriers) connect to low band antennas.
- Mid bands radios (carriers) connect to mid band antennas.
- High band radios (carriers) connect to high band antennas, etc.

Following are the examples of the resulting radio (carrier) numbers and antenna configuration per site. There can be multiple radios (carriers) per sector for each technology.

For example:

- Two 3G radios (carriers) per sector (and multiple 2G radios per sector)
- Three 4G radios (carriers) per sector
- Two 5G radio (carrier) per sector.

Each radio (carrier) connects to an antenna and antennas contained within antenna panels of a sector. The number of radios (carriers) for each technology varies for each site, varies with traffic levels, RAN design, technology deployment, etc. The number of radios (carriers) for each site is information provided by the mobile operator, but now you should be able to understand RF/RAN references to carriers, sectors etc.

4.2.2 Mobile network spectrum

The mobile network spectrum allocation for 2G, 3G, 4G, and 5G networks can vary across different regions and countries. However, here is a general overview of the frequency bands commonly used for each generation (Figure 4.2):

2G (second generation): 2G networks, such as GSM (Global System for Mobile Communications), primarily operate in the 900 MHz and 1800 MHz frequency bands. Some regions also utilize the 850 MHz and 1900 MHz bands, particularly in North America.

Figure 4.2: Radio frequency spectrum.

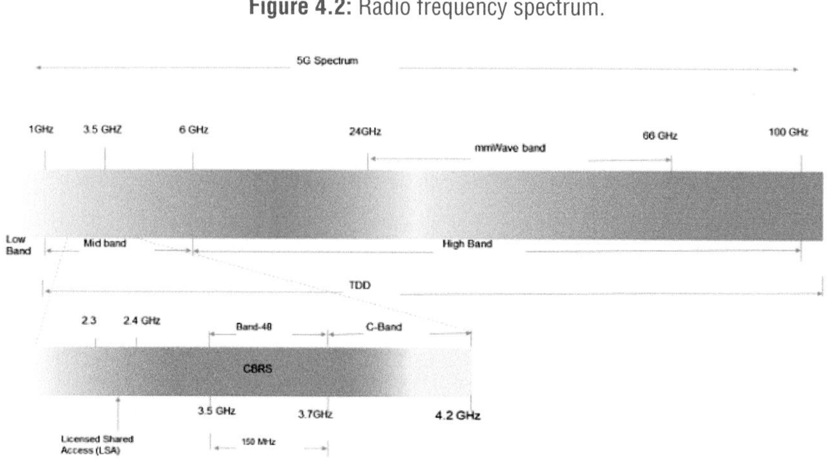

Ref: O-RAN.WG4.CUS.0-v05.00. O-RAN Fronthaul Working Group

3G (third generation): 3G networks, including technologies like UMTS (Universal Mobile Telecommunications System) and CDMA2000, typically utilize frequency bands around 850 MHz, 900 MHz, 1900 MHz, and 2100 MHz. These frequency bands can vary depending on regional allocations and specific technology deployments.

4G (fourth generation): 4G networks, based on LTE (long-term evolution) technology, operate in a wide range of frequency bands. Some of the commonly used bands include 700 MHz, 800 MHz, 900 MHz, 1800 MHz, 2100 MHz, 2300 MHz, 2500 MHz, and 2600 MHz. The specific band allocations can differ among countries and regions.

5G (fifth generation): 5G networks introduce a wider range of frequency bands to support higher data rates and lower latency. The key frequency bands for 5G include sub-6 GHz bands (e.g., 600 MHz, 700 MHz, 2.5 GHz, 3.5 GHz, and 3.7–4.2 GHz), as well as mmWave (millimeter-wave) bands in the 24 GHz, 28 GHz, and 39 GHz ranges. The specific frequency bands deployed for 5G can depend on regulatory approvals and regional allocations.

As shown in Figure 4.3, it is important to note that more spectrum means more capacity and hence more network bandwidth. And lower frequency bands provide better propagation. The spectrum is further subdivided into channels, e.g. 5 MHz, 10 MHz, 20 MHz (RF) bandwidth channels. Wider (bigger) channels allow for more data; higher bands have capacity for more channels.

Figure 4.3: 5G radio frequency bands – coverage vs. throughput.

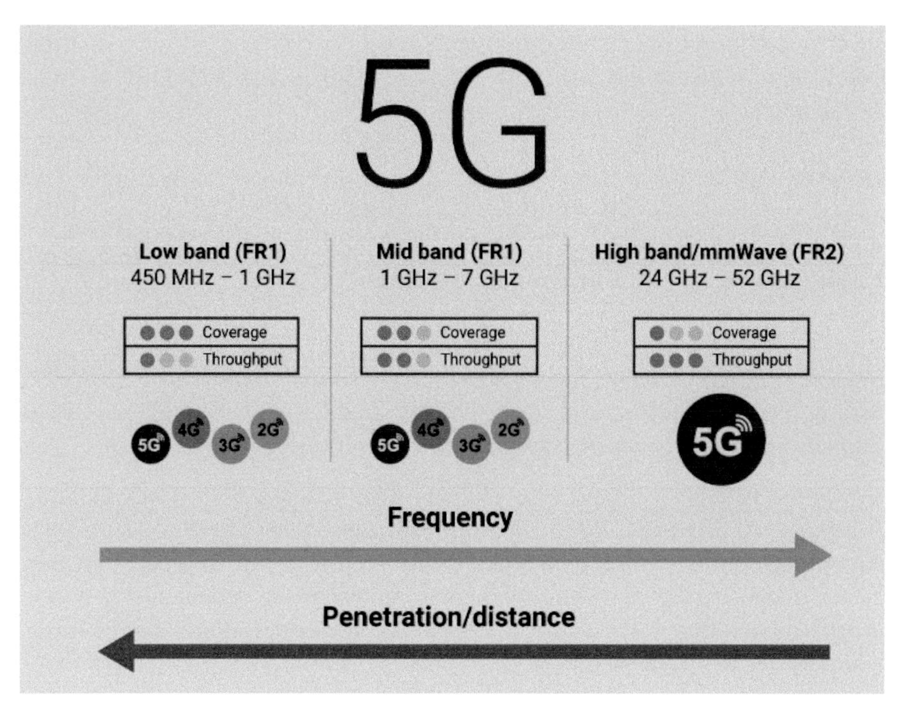

These frequency bands are not exclusive to a specific generation, and some bands may be used for multiple generations depending on network deployments and technology advancements. Additionally, the spectrum allocation can vary between different countries and regions based on regulatory policies and availability of frequencies.

4.2.3 MIMO (multiple input, multiple output)

MIMO is a technology which helps to improve data throughput, increase network capacity, and enhance signal quality. MIMO utilizes multiple antennas at both the transmitter and receiver to transmit and receive multiple data streams simultaneously. In MIMO systems, the multiple antennas at the transmitter and receiver enable spatial multiplexing, diversity, and beamforming techniques:

Spatial multiplexing: MIMO takes advantage of the spatial dimension to transmit multiple data streams simultaneously over the same frequency band. Each antenna transmits a different data stream, and the receiver uses multiple antennas to separate and decode these streams, effectively increasing the data throughput.

Diversity: MIMO also provides diversity by transmitting the same data stream across multiple antennas. This improves signal reliability by mitigating the effects of fading and interference. The receiver combines the signals from multiple antennas to enhance the received signal quality.

Beamforming: With MIMO, it is possible to shape and direct the wireless signals towards specific locations. By adjusting the phase and amplitude of the signals transmitted from different antennas, MIMO enables beamforming, which can improve coverage, signal strength, and overall network performance.

MIMO technology is used in various wireless communication standards, including 4G LTE, 5G, Wi-Fi (802.11n, 802.11ac, and 802.11ax), and other wireless systems. The number of antennas used in MIMO systems can vary, ranging from 2 × 2 (two transmit antennas and two receive antennas) to higher configurations like 4 × 4 or 8 × 8, depending on the specific implementation and network requirements.

The multiple data streams at each end of the radio communications link are combined to minimize errors and optimize data speed. MIMO can increase the number of data streams between base station radio and user equipment, thereby increasing the data speed and/or the reliability. MIMO can increase the overall data throughput for a radio base station. When MIMO increases the data throughput for a radio base station it impacts the xHaul (more specifically fronthaul) dimensioning.

By utilizing multiple antennas and advanced signal processing techniques, MIMO technology enhances the capacity, coverage, and reliability of wireless communication systems, enabling higher data rates and better overall performance.

In Figure 4.4, the user equipment (UE) is in good coverage signal strength conditions where four layers are employed for four separate simultaneous data stream connections (four layers) between 4G radio (carrier) and UE. The result is a 4× increase in throughput for the UE.

Figure 4.5 shows that the user equipment (UE) is in average/weak coverage signal strength conditions (cell edge). Four layers are employed for two separate simultaneous data stream connections (using four layers) between 4G radio

Figure 4.4: 4 × 4 MIMO to increase bandwidth.

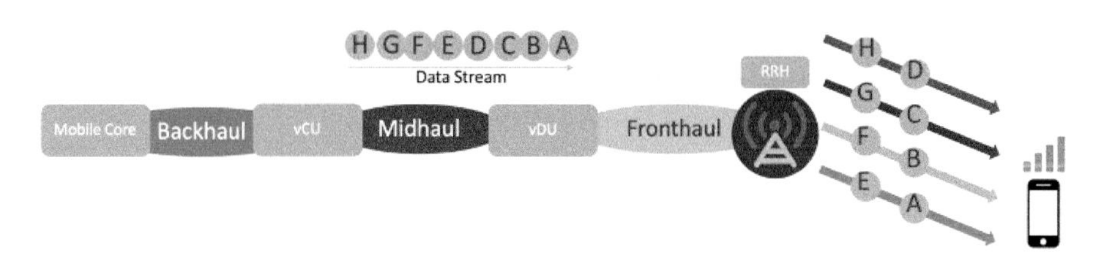

Figure 4.5: 4 × 4 MIMO to increase reliability.

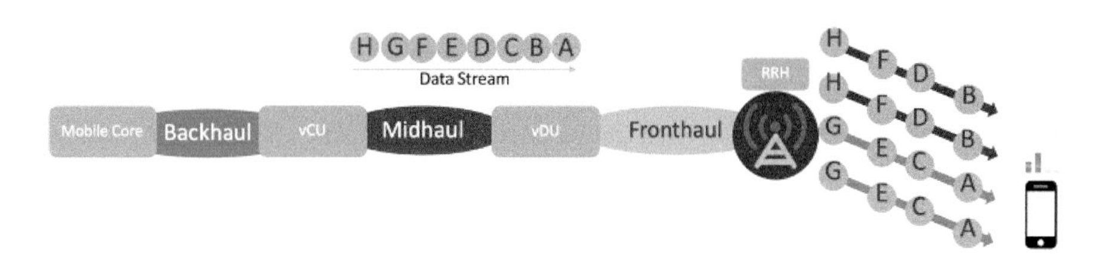

(carrier) and UE and each data stream duplicated on two separate layers. The result is a two times increase in throughput for the UE and greater reliability/performance of the connection to a UE at the cell edge.

4.3 Backhaul Evolution to xHaul to Support 5G Use Cases

While backhaul networks play a critical role in supporting 5G networks, they also have some limitations that need to be addressed to fully realize the potential of 5G. Here are a few key limitations:

Capacity: 5G networks are designed to handle significantly higher data rates and traffic volumes compared to previous generations. However, traditional backhaul networks may not have the capacity to support the increased bandwidth requirements of 5G. Upgrading backhaul links, such as transitioning from legacy TDM to higher-capacity Ethernet or optical fiber, is essential to address this limitation.

Latency: 5G promises ultra-low latency for applications like autonomous vehicles, remote surgery, and real-time gaming. However, backhaul networks may introduce additional latency due to the distance between base stations and core networks. Traffic traverses all of service provider aggregation and core leading to unnecessary bandwidth use in the provider network. The problem will only grow with high bandwidth usage in 5G. Reducing backhaul latency through technologies like edge computing and localized data centers can help overcome this limitation.

Scalability: With a backhaul network in D-RAN the cell sites are heavy because each cell site has antennas, a radio unit (RU) or remote radio head (RRH) and baseband unit (BBU), which also leads to real estate problems. The evolution to centralized RAN (C-RAN) with a xHaul network solves this problem as the BBUs are centralized. The massive scale of 5G networks, with a significantly higher number of connected devices and network elements, poses scalability challenges for backhaul networks. Scaling the backhaul infrastructure, including capacity, connectivity, and management capabilities, is crucial to support the increasing number of base stations and devices in 5G networks.

Synchronization: 5G networks require precise synchronization to ensure coordinated operation among base stations and support advanced features like beamforming. Traditional backhaul technologies, such as Ethernet, may have limitations in providing the required synchronization accuracy. Implementing synchronization mechanisms, such as the IEEE 1588 Precision Time Protocol (PTP), is necessary to address this limitation.

Deployment challenges: Deploying a new backhaul infrastructure, such as laying optical fiber or establishing microwave links, can be complex and time-consuming. The availability of suitable sites, rights-of-way, and regulatory approvals can pose challenges in expanding backhaul networks to support 5G deployments, especially in remote or underserved areas.

Addressing these limitations requires significant investment, planning, and coordination among mobile operators, backhaul providers, and regulatory bodies. Upgrading and expanding backhaul networks to meet the capacity, latency, scalability, synchronization, and deployment challenges of 5G is crucial to ensure the successful implementation and performance of next-generation mobile networks.

Backhaul evolution to xHaul refers to the transformation of traditional backhaul networks in mobile communication systems to a more flexible, efficient, and converged transport infrastructure known as "xHaul" (pronounced as “cross-haul“). This evolution is driven by the increasing demands of 5G networks and the need to support diverse services and applications with high throughput, low latency, and stringent quality of service (QoS) requirements.

Historically, backhaul networks were designed to transport traffic from base stations (eNodeBs in LTE networks) to the core network. They primarily utilized technologies like TDM (time division multiplexing), Ethernet, or microwave links. However, with the advent of 5G and the proliferation of various access technologies, the traditional backhaul networks were not sufficient to meet the demands of the evolving mobile landscape.

xHaul represents a paradigm shift in backhaul networks, as it aims to converge multiple types of transport networks into a unified and flexible architecture. It combines traditional backhaul with fronthaul (connecting baseband units to remote radio heads) and midhaul (connecting distributed edge data centers). xHaul networks support various transport technologies, including optical fiber, Ethernet, IP/MPLS (internet protocol/multi-protocol label switching), and even wireless links like microwave and millimeter-wave.

The transition from a distributed radio access network (D-RAN) to a centralized radio access network (C-RAN) involves a shift in the architecture and deployment of radio access networks. Here are some key considerations and steps involved in this transition:

BBU centralization: In D-RAN, the baseband units (BBUs) are distributed closer to the remote radio heads (RRHs) at the cell sites. In the transition to C-RAN, the BBUs are centralized in a central location or data center. This consolidation of BBUs allows for more efficient resource allocation, easier management, and potential cost savings.

High-speed fronthaul: C-RAN relies on high-speed and low-latency fronthaul connections between the centralized BBUs and RRHs. Fronthaul refers to the link that carries baseband signals between the BBUs and RRHs. Upgrading or deploying high-capacity and low-latency fronthaul links, such as optical fiber or Ethernet, is essential for a successful transition to C-RAN.

Virtualization and cloud technologies: C-RAN often involves the virtualization of network functions, where baseband processing functions are implemented as software running on standard servers. Cloud technologies, such as network function virtualization (NFV) and software-defined networking (SDN), are utilized to enable the centralized control and management of the RAN resources.

Centralized baseband processing: In C-RAN, the centralized BBUs handle the baseband processing functions for multiple RRHs. This centralized processing allows for better coordination, resource pooling, and advanced features like coordinated multi-point (CoMP) transmission and reception.

Synchronization: Synchronization becomes even more critical in C-RAN due to the centralized processing and coordinated operation. Precise synchronization mechanisms, such as the IEEE 1588 Precision Time Protocol (PTP) or synchronous Ethernet, are used to ensure accurate timing and synchronization between the BBUs and RRHs.

Capacity and scalability: C-RAN offers increased capacity and scalability by allowing for dynamic allocation of resources based on network demand. The centralized architecture enables efficient resource utilization and easier scaling of the network to accommodate growing traffic volumes and future technologies like 5G.

While C-RAN offers numerous benefits, such as improved performance, resource efficiency, and cost savings, the transition from D-RAN to C-RAN requires careful planning, infrastructure upgrades, and coordination among network operators, equipment vendors, and service providers.

The key objectives of the backhaul evolution to xHaul are:

Flexibility: xHaul networks are designed to support various access technologies, including 5G, LTE, and even legacy 2G/3G systems. They provide the flexibility to accommodate different bandwidth requirements, radio interfaces, and deployment scenarios.

Scalability: xHaul networks are scalable to handle the increased traffic volumes and the massive number of connected devices expected in 5G networks. They can efficiently handle the anticipated growth in network capacity and the number of connections.

Low latency: xHaul architecture is optimized for low-latency transport to support ultra-reliable and low-latency communication (URLLC) services that require minimal delay, such as autonomous vehicles, remote surgery, and industrial automation.

Convergence: xHaul networks converge multiple types of traffic onto a single transport infrastructure, reducing complexity and operational costs. This convergence enables efficient resource utilization and end-to-end network management.

The evolution from traditional backhaul to xhaul is crucial for enabling the full potential of 5G networks, supporting emerging services and applications, and providing an efficient and flexible transport infrastructure for mobile operators.

C-RAN architecture and components:

Centralized radio access network (C-RAN) is an architecture that centralizes the baseband processing functions of a radio access network, enabling efficient resource sharing and coordination among multiple remote radio heads (RRHs). The key components of a typical C-RAN architecture include the baseband unit (BBU), remote radio head (RRH), fronthaul links, fronthaul gateways, and a centralized controller.

The baseband unit (BBU) is responsible for processing the baseband signals, including modulation, coding, and beamforming. In C-RAN, the BBUs are centralized in a central location or data center, rather than being distributed to each cell site. By centralizing the BBUs, resource sharing and coordination become possible, leading to improved efficiency and cost savings.

The remote radio head (RRH) is the radio transceiver unit located at the cell sites. It is responsible for transmitting and receiving radio signals to and from user devices. In a C-RAN architecture, the RRHs are connected to the centralized BBUs through high-speed fronthaul links. These fronthaul links carry the digitized baseband signals between the BBUs and RRHs, ensuring efficient communication between them.

The fronthaul links can be implemented using various technologies, such as optical fiber, Ethernet, or even wireless solutions like microwave or millimeter-wave. The fronthaul links must have sufficient capacity and low latency to accommodate the high-speed transmission of baseband signals.

To interface between the centralized BBU pool and the RRHs, a fronthaul gateway is employed. The fronthaul gateway receives the baseband signals from the BBUs and converts them into the appropriate format for transmission over the fronthaul links. Additionally, it may perform functions like packetization, error correction, and synchronization distribution to ensure the proper delivery of signals between the BBUs and RRHs.

A centralized controller, often referred to as the central unit (CU), is responsible for managing and controlling the operation of the BBUs and RRHs in the C-RAN architecture (Figures 4.6 and 4.7). The centralized controller coordinates the allocation of radio resources, manages interference coordination, and facilitates advanced features like coordinated multi-point

(CoMP) transmission and reception. It plays a crucial role in optimizing network performance, enhancing capacity, and improving the overall efficiency of the C-RAN architecture.

Figure 4.6: Centralized radio access network (C-RAN).

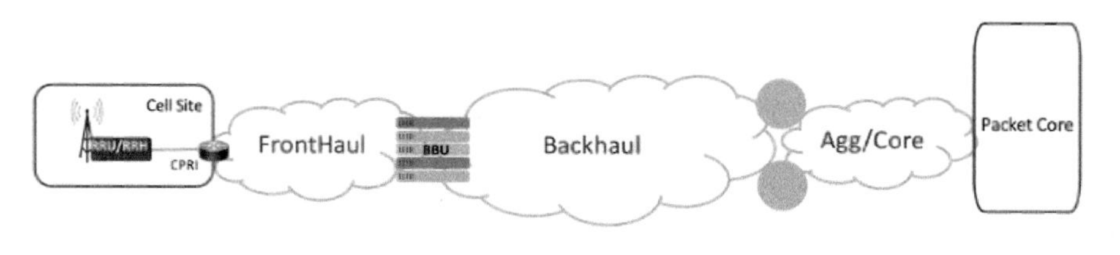

Figure 4.7: C-RAN BBU split with multi-domain orchestration.

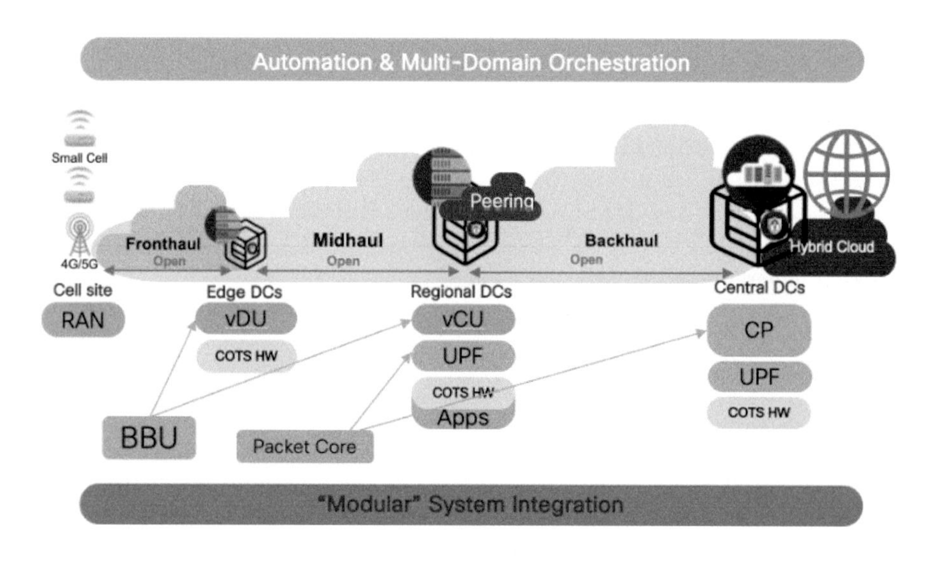

The CPRI is the optical interface between RU/RRU and BBU. In proprietary RAN vendor format, typically the RRU and BBU are from same vendor. There is traffic sent at a constant bit rate even if there is no customer data. This is a very high bit rate ~2.5G, whereas the BBU-CSR bit rate is typically ~300 Mbps or less. The network between RRU and BBU is called the CPRI fronthaul. When multiple BBU's are placed in a central location, it helps to save on real estate requirements at every cell site (Figure 4.8).

Figure 4.8: C-RAN hub site.

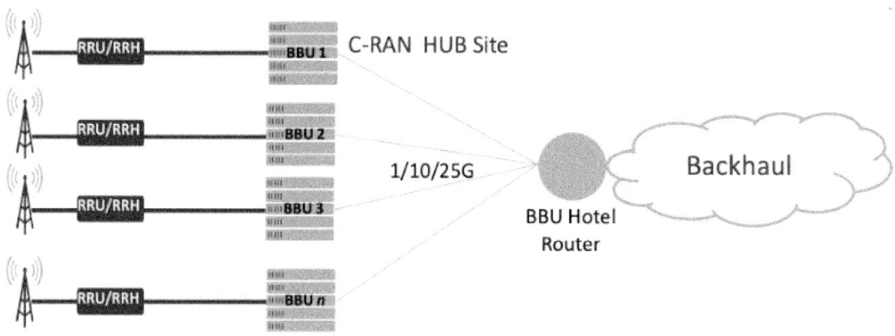

Multiple BBU's can be placed in the central location, but there is a 1:1 relationship between the cell site/RRU and BBU. The C-RAN hub site could be a shelter site that has space and power for multiple BBUs. It helps to achieve real-estate savings, i.e. multiple street light antennas could go to the same C-RAN hub. There are low latency requirements for RRU-BBU links; typically <200 μs but it depends on the RAN vendor. A BBU hotel router aggregates BBUs for transport to mobile backhaul. This is feasible for fiber rich environments for CPRI transport.

Overall, the C-RAN architecture with its components enables resource pooling, centralized management, and coordination in the radio access network, resulting in improved efficiency, reduced costs, and enhanced network performance.

4.4 CPRI Functional Splits and Evolution to eCPRI

As explained in the previous section, C-RAN saves on real-estate, but still requires the same number of BBUs, uses the same rack space in a central location, the same power requirements but within a specialized (closed) piece of hardware.

Virtualization can be further used to optimize real-estate, power and rack space which will result in a virtualized BBU (vBBU) and could be run on a cluster of COTS hardware. This allows for further "decomposition" of RAN by introducing "functional splits" of CPRI. BBU functionality may be split into a distributed unit (DU) and centralized unit (CU). DU needs near real-time communication with RU, i.e. very strict latency requirements in the packetized fronthaul network.

eCPRI (enhanced common public radio interface) is a protocol used in the fronthaul network of a virtualized RAN (V-RAN) architecture (Figure 4.9). It defines different functional splits between the centralized unit (CU) and the distributed unit (DU) in order to optimize the allocation of processing tasks and minimize the fronthaul bandwidth requirements.

Figure 4.9: CPRI to eCPRI evolution.

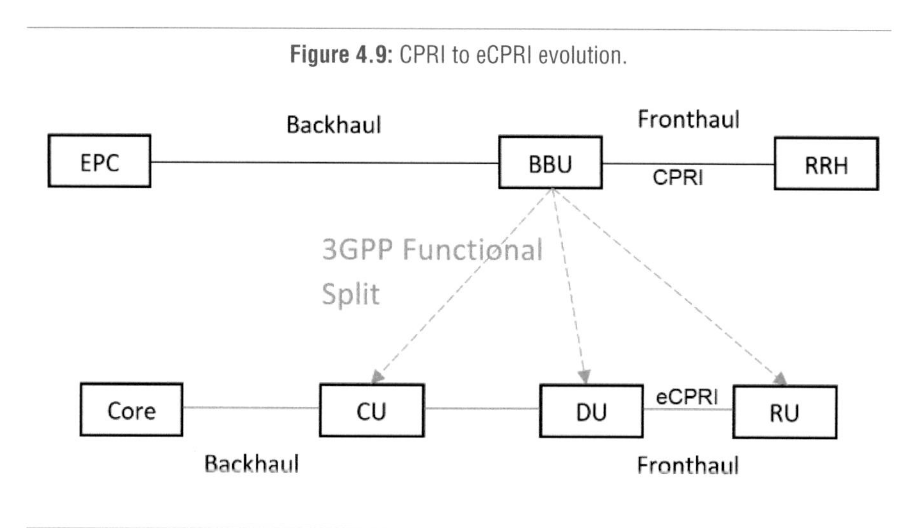

Radio over Ethernet (RoE) mappers are used to carry radio traffic over packet network (Figure 4.10). A structure agnostic RoE mapper is a function that converts other transport framing formats to a RoE framing format, and a RoE de-mapper performs the opposite function (Figure 4.11). This mapper captures bits from one end of a constant bit rate link, packetizes the bits into ethernet frames, sends the packets across the network, and then recreates the bit stream at far end of the link. There are two types:

Type-0: This works as a simple Ethernet tunnel. It does not remove any line coding bits and doesn't interpret any special characters. If source-data is 8B/10B encoded, the 10-bit symbols present on the line will be tunneled by the mapper as 10 bits data.

Type-1: This works in line coding aware mode. It removes line codes and adds them back. If source data is 8B/10B encoded, after the decoding process, the 8-bit symbols present on the line will be tunneled by the mapper as 8-bits. On the

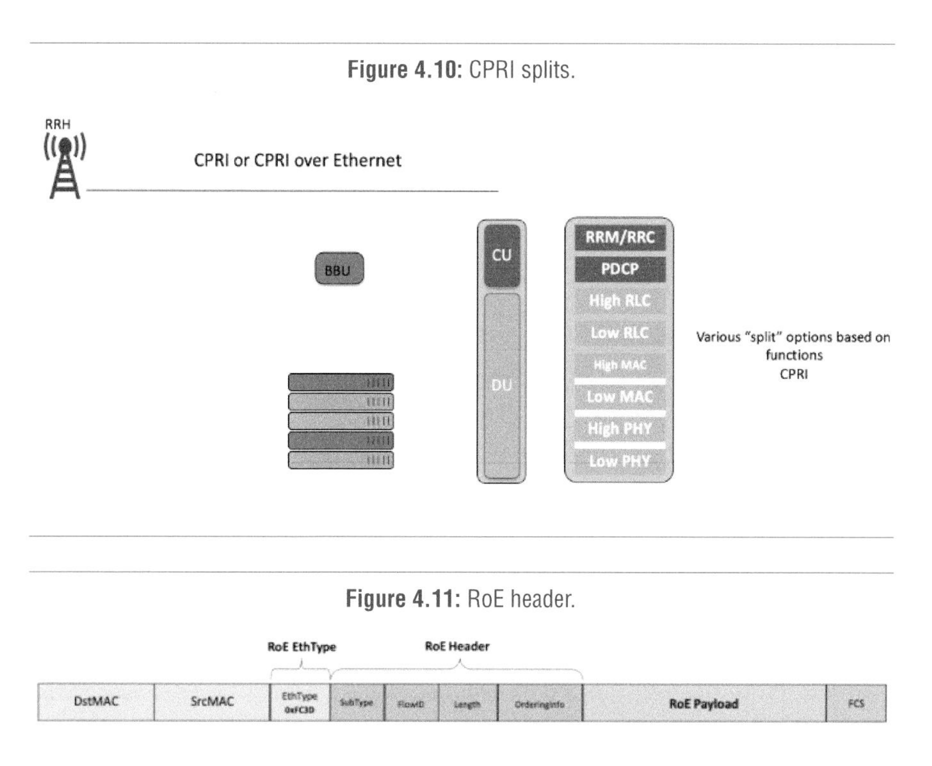

Figure 4.10: CPRI splits.

Figure 4.11: RoE header.

other hand, the 8-Bit will be encoded back to 10-bits using a standard coding-scheme.

Figure 4.12 shows CPRI splits with latency and bandwidth and is composed by taking a reference from this 3GPP link https://www.3gpp.org/news-events/3gpp-news/open-ran. Here is a detailed explanation of eCPRI functional splits:

eCPRI functional split option 2: In this split, the CU and DU share the processing tasks. The CU is responsible for high-layer processing functions, including radio resource management, scheduling, and network management. The DU takes care of lower-layer processing tasks, such as radio signal processing, encoding/decoding, and modulation/demodulation. This split strikes a balance between the CU and DU, ensuring efficient utilization of processing resources and reducing the fronthaul bandwidth required to transmit data between them.

eCPRI functional split option 3: This split further offloads processing tasks from the DU to the CU. In addition to the high-layer functions, the CU now

Figure 4.12: CPRI splits with latency and bandwidth.

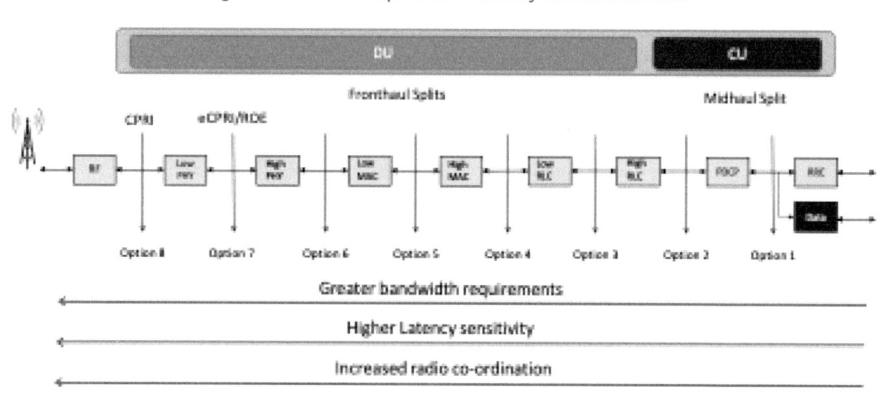

also handles lower-layer tasks like digital beamforming and precoding. The DU primarily focuses on radio signal processing and other lower-layer functions. By centralizing more processing tasks in the CU, this split reduces the complexity and cost of the DU, which can now be implemented as simpler hardware units.

eCPRI functional split option 4: This split maximizes the centralization of processing tasks in the CU. The CU takes on all lower-layer processing functions, including radio signal processing, encoding/decoding, modulation/demodulation, and digital beamforming. The DU in this split acts primarily as a radio signal transceiver, converting the analog radio signals to digital and transmitting them to the CU over the fronthaul network. This split significantly simplifies the DU, enabling cost-effective deployment of V-RANs.

The transport costs are minimized with higher splits, i.e. as we move from option 8 to option 1. Radio frequency (RF) gains are improved with lower splits, i.e. as we move from option 1 to option 8.

Each eCPRI functional split offers trade-offs between fronthaul bandwidth requirements, processing complexity, and resource utilization. Option 2 provides a balanced approach, suitable for scenarios where fronthaul capacity is limited, and there is a need for distributed processing capabilities. Option 3 offloads more processing tasks to the CU, reducing the complexity and cost of the DU. Option 4 centralizes most processing tasks in the CU, simplifying the DU but increasing fronthaul bandwidth requirements.

The choice of eCPRI functional split depends on factors such as network conditions, deployment scenarios, and resource availability. For example, in deployments with limited fronthaul capacity, option 2 might be preferred to minimize bandwidth requirements. In scenarios where cost and complexity reduction are important, option 4 could be chosen to centralize processing tasks in the CU.

The eCPRI functional splits enable flexibility and scalability in V-RAN deployments. Operators can adapt the functional split based on changing network conditions, traffic demands, or evolving requirements. The ability to adjust the allocation of processing tasks between the CU and the DU allows for optimized resource utilization and performance.

Overall, the eCPRI functional splits provide a framework for efficiently distributing processing tasks in V-RAN architectures. They help optimize resource utilization, reduce costs, and minimize fronthaul bandwidth requirements, ultimately enhancing the performance and scalability of virtualized RAN deployments.

There are no ways to connect the proprietary CPRI RAN interface to a packet router and that's why the RAN fronthaul standards are evolving towards open RAN (O-RAN) (Figure 4.13). An operator led RAN alliance to create an open ecosystem around RAN. There is no more vendor lock-in for RU-BBU(DU/CU) interfaces which makes it easier for network equipment vendors to interoperate with RAN equipment. Presently there are 200+ operators and vendors that are part of the alliance and they actively participate in multiple working groups.

Figure 4.13: Moving towards O-RAN.

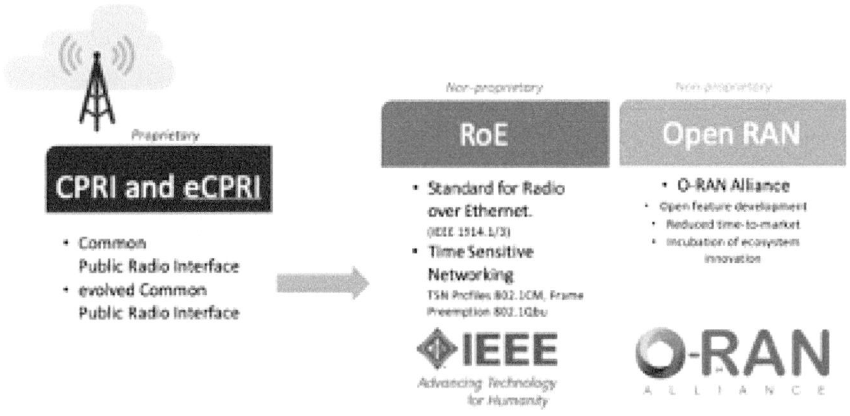

4.5 Fronthaul Design Considerations

Designing the fronthaul network is a critical consideration in a virtualized radio access network (V-RAN) architecture for 5G networks (Figure 4.14). The fronthaul network connects the virtual centralized unit (vCU) and virtual distributed unit (vDU) components, carrying digitized radio signals, control signals, and synchronization information.

Figure 4.14: Fronthaul considerations.

Here are some key design considerations for fronthaul in V-RAN for 5G networks:

Transport network: The transport network is the underlying infrastructure that carries the fronthaul traffic to the vDU and further to the vCU via the midhaul. The vCU is responsible for high-level processing functions, such as radio resource management, scheduling, and network management. The vDU handles lower-level processing tasks, including radio signal processing, encoding/decoding, and modulation/demodulation. The fronthaul design ensures seamless communication and coordination between the vCU and vDU to maintain efficient operation and optimize resource utilization. It can be based on various technologies, including fiber optics, Ethernet, or wireless links. The transport network must have sufficient capacity, low latency, and high reliability to support the high-bandwidth requirements and low-latency demands of 5G networks.

Bandwidth and latency: 5G networks require high bandwidth and low latency fronthaul connections to support the massive increase in data traffic and the stringent performance requirements of emerging applications like augmented reality, virtual reality, and autonomous vehicles. The fronthaul network must be designed to accommodate the high-speed transmission of large volumes of data with minimal delay.

Network slicing: Network slicing is a key feature of 5G networks, allowing operators to create multiple virtual networks on a shared physical infrastructure to cater to different use cases and customer requirements. The fronthaul design should support efficient and flexible allocation of network resources for different slices, ensuring the required quality of service and isolation between slices.

Synchronization: Accurate synchronization is crucial in 5G networks to support advanced features like coordinated multi-point (CoMP) transmission and beamforming. The fronthaul network must provide precise synchronization capabilities, such as distributing time and frequency references, to ensure synchronized operation between the CU and DU components.

Fronthaul capacity: The fronthaul network should have sufficient capacity to handle the increased traffic demands of 5G networks. The higher number of antennas and wider bandwidths in 5G systems result in increased fronthaul data rates. Designing the fronthaul network with adequate capacity and scalability is essential to support future growth and accommodate the evolving needs of 5G deployments.

Protocol considerations: The choice of fronthaul protocol is crucial in V-RAN for 5G networks. Protocols such as eCPRI (enhanced common public radio interface) and IEEE 1914.x provide efficient transport of digitized radio signals and control information with low latency and high bandwidth efficiency. The fronthaul design should consider the protocol selection based on factors like bandwidth requirements, latency tolerance, and compatibility with the virtualized architecture.

Scalability and future-readiness: The fronthaul design should be scalable and future-ready to accommodate the evolving needs of 5G networks. It should be able to handle increasing traffic demands, support higher bandwidths, and adapt to new technologies and protocols. Designing a flexible and scalable fronthaul network ensures that it can grow and evolve with the network requirements, reducing the need for costly and disruptive upgrades in the future.

Overall, designing the fronthaul network in V-RAN for 5G networks involves careful consideration of bandwidth, latency, network slicing, synchronization, capacity, and protocol requirements. A well-designed fronthaul network ensures efficient and reliable transmission of radio signals, supports the advanced features of 5G networks, and enables the seamless integration of virtualized components in the RAN architecture.

4.6 O-RAN xHaul Architecture

The concept of 5G fronthaul decomposition is further democratized by O-RAN (open RAN) into composable hardware and software encouraging wider industry participation. The O-RAN alliance (https://www.o-ran.org/) specified operational and configuration standards extending the concept of 3GPP NG-RAN architecture. The idea advocated by OpenRAN for this notion of decomposition is to break the shackle of vendor locked systems to vendor agnostic systems while bringing in the collective innovation of industry to endow better solutions for 5G fronthaul. In O-RAN/OpenRAN, RF and lower PHY of the radio protocol stack is processed in a radio unit (RU) and the upper PHY to RLC (radio link control) is processed in a DU (distributed unit); much of the packetization starting at PDCP (packet data convergence protocol) to layer 3 encapsulation are done at the CU (central unit). In the O-RAN/OpenRAN model of RAN infrastructure, RU is known as O-RU, DU is known as O-DU and CU is known as O-CU.

This concept of RAN infrastructure decomposition is well received by the industry and there is an increase in the deployment of the O-RAN fronthaul infrastructure concept. RU is connected to DU over a fiber link for which the underlay is Ethernet whether it is for eCPRI or RoE (radio over Ethernet). Furthermore, DU to CU and CU to 4G EPC or 5G core are also connected through the Ethernet link. The OpenRAN model pretty much adopted the concept of “open networking” for RU, DU and CU. In it, a whitebox type of “off the shelf” hardware is used for RU, DU and CU solutions. More specifically, DU and CU utilize standard “off the shelf” servers while the RU may or may not use “off the shelf” servers. Irrespective of how the deployments are done, synchronization remains a critical imperative of the network configuration.

The concept of OpenRAN can be further understood considering how 3GPP 5G NG-RAN architecture suggests the split of the radio protocol stack as depicted in Figure 4.15.

Figure 4.15 shows the concept of merging of NG-RAN and O-RAN to show how the radio protocol stack is split to achieve network decomposition. The most common split is 7:2 in which RF and LPHY (lower PHY) of the radio protocol stack remain in the radio unit and UPHY (upper PHY) to URLC (upper radio link control) are processed within the DU (distributed unit). The PDCP (packet data convergence protocol) function is performed at the CU (central unit).

Figure 4.15: Radio protocol stack split to achieve network decomposition.

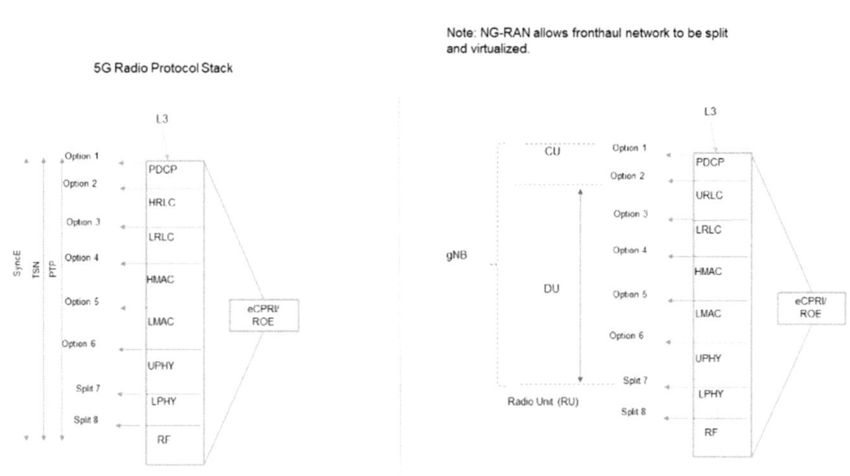

Ref: O-RAN, 2021. Control, User and Synchronization Plane Specification: O-RAN.WG4.CUS.0-v05.00. O-RAN Fronthaul Working Group.

As suggested in the diagram, the entire split end-to-end requires synchronization meaning RU to CU; the designer must consider highly precise synchronization. Recently, TIP specified the need for synchronization from DU to CU and suggested the deployment of SyncE between DU and CU.

The DU may be able to implement a grandmaster or boundary clock while the RU implements a slave clock. A GNSS timing module either single band (ICM 360) or dual band (RES720) can be inserted to RU and DUs. For even better solutions, GNSS timing capabilities and PTP can be brought in a single embedded board known as a grandmaster clock. If the configuration of DU requires an external clocking device, then grandmaster offers two-in-one solutions: boundary clock and grandmaster. Please note, all OpenRAN deployment must consider ITU-T guidelines for an end-to-end time error budget of less than 1.1 μs (Figure 4.16).

There is an increased momentum towards adoption of the OpenRAN concept as although implementation may vary, much of the OpenRAN design construct remains the same. In it, 3 to 9 RUs are connected per DU and the DUs are connected either directly or through a switch known as a CSR (cell site router) or DCSG (disaggregated cell site gateway) to CU.

Figure 4.16: Typical O-RAN deployment.

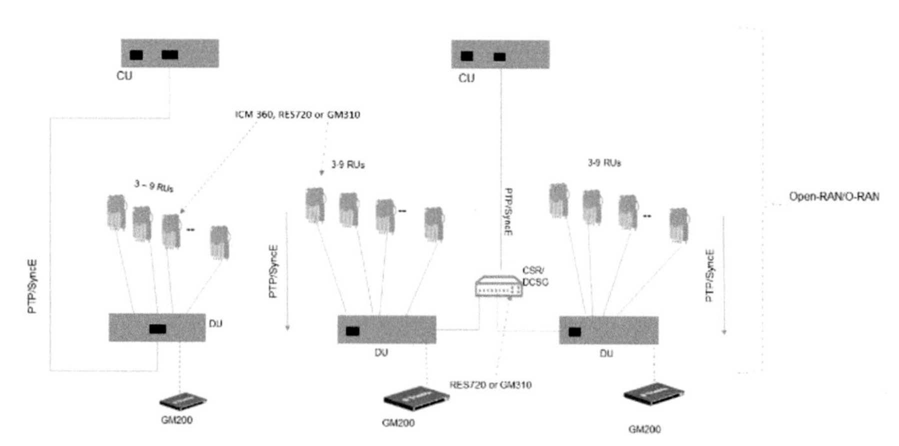

Ref: O-RAN, 2021. Control, User and Synchronization Plane Specification: O-RAN.WG4.CUS.0-v05.00. O-RAN Fronthaul Working Group

CHAPTER

5

IP Transport: xHaul Architecture Design

5.1 Introduction

The IP xHaul transport architecture refers to the network infrastructure that enables the transport of IP-based traffic between the radio access network (RAN) and the core network in a mobile network environment. IP xHaul transport provides connectivity for various generations of mobile networks, including 3G, 4G, and 5G.

Overall, the IP transport architecture provides the foundation for the internet and enables the seamless communication and exchange of data between devices and networks worldwide.

3G and 4G are generations of mobile communication technologies that brought significant advancements in data transmission speeds and capabilities. The transport networks supporting these technologies are designed to handle the unique requirements and characteristics of 3G and 4G mobile networks. Here's an overview of the transport networks associated with 3G and 4G:

3G transport network:

- Circuit-switched core: 3G networks initially relied on circuit-switched networks for voice calls. These networks were primarily based on legacy technologies such as TDM (time division multiplexing) and ATM (asynchronous transfer mode).
- Packet-switched core: With the growth of data services, 3G networks introduced a packet-switched core network called the UMTS (universal mobile telecommunications system) core. It utilized protocols such as IP (internet protocol) and ATM to handle data traffic.

- Backhaul network: The backhaul network connects the base stations (Node B in UMTS) to the core network. It often used technologies like T1/E1 lines, microwave links, or DSL (digital subscriber line) connections to transport traffic from the base stations to the core network.

4G transport network:

- IP-based core: 4G networks transitioned to a fully IP-based core network architecture known as the evolved packet core (EPC). This allowed for a more efficient and streamlined transport of both voice and data traffic.
- Packet-switched core: The EPC is based on packet-switching technologies, utilizing IP protocols throughout the core network. It offers increased flexibility, scalability, and support for advanced services and applications.
- Ethernet backhaul: 4G networks typically employ Ethernet-based backhaul for connecting the base stations (eNodeBs in LTE) to the core network. Ethernet provides higher bandwidth, lower latency, and easier management compared to previous backhaul technologies.
- Fiber and microwave links: To support the increased data capacity and speed requirements of 4G, fiber-optic links are often used for backhaul connections. In areas where fiber is not available, microwave links can be used as a wireless backhaul solution.

3G and 4G transport networks involve a combination of wired and wireless technologies, with a focus on packet-switched networks and IP protocols. The transition to an all-IP architecture in 4G enables faster data rates, lower latency, and more efficient utilization of network resources. These advancements laid the foundation for the subsequent evolution to 5G networks, which offer even higher speeds, lower latency, and enhanced capabilities for diverse applications.

5G will have a wider wireless spectrum, with a higher frequency band; the bandwidth for 5G networks can even reach tens of Gbps. The peak bandwidth and user experience bandwidth of 5G networks is expected to be 10 times higher. 5G comes with its own benefits, but it also imposes some challenges on the existing telecom transport network. The existing technical infrastructure including the equipment and the network architectures does not fulfill the requirements of deploying a 5G network.

While addressing these challenges it is also important that an optimal cost solution is considered for successful deployment and rollout of commercial 5G services to meet consumer demands at the same time as securing market leadership with sustainable growth. How can this be achieved is a question most of the telecom operators are prioritizing and trying to address with their 5G deployment plans.

IP backbone, metro and mobile backhaul are the three main aspects of a transport IP network. Improving the network efficiency by leveraging new technologies such as SDN and programmability has become of utmost importance. Telecom operators today are looking for network operators who can provide an IP transport network, which is capable of building future-oriented ultra-broadband, low-latency reliable and high-quality, multi-service bearing network to help operators reduce investment and operation costs, improve end-user experience, and support rapid business development.

5.2 Present IP Transport Architecture

The end user traffic has increased in leaps and bounds over the last decade and so the transport networks went through evolutions to cater for the growing scale demands and bandwidth requirements. Operators are looking for robust scale with operational simplicity and resiliency. In recent years, transport networks have evolved from traditional plain IP based transport to MPLS-LDP to SR-MPLS.

In 3G and 4G networks, multiprotocol label switching (MPLS) technology is commonly used for transport to provide efficient and reliable packet forwarding. MPLS offers several benefits in terms of traffic engineering, quality of service (QoS), and scalability. Here's an overview of MPLS transport in 3G and 4G networks:

- Label switched paths (LSPs): In an MPLS network, data packets are forwarded based on labels. LSPs are established between the network nodes to define the path that packets will follow. LSPs are unidirectional and can be set up dynamically or through manual configuration.
- Label distribution: The MPLS network uses a signaling protocol such as the label distribution protocol (LDP) or resource reservation protocol (RSVP) to distribute labels and establish LSPs between network nodes. These protocols enable the nodes to exchange information and establish the appropriate forwarding paths.
- Traffic engineering: MPLS provides traffic engineering capabilities, allowing network operators to control the path that traffic takes through the network. With MPLS, operators can optimize network resource utilization, balance traffic load, and prioritize traffic based on QoS requirements.
- Quality of service (QoS): MPLS supports QoS mechanisms by allowing the classification and prioritization of traffic flows based on different criteria. By assigning appropriate labels and implementing QoS policies, MPLS can ensure the delivery of real-time and latency-sensitive traffic with specific bandwidth and latency requirements.
- MPLS-based backhaul: In 3G and 4G networks, MPLS is often used for the backhaul transport of traffic from the base stations (Node Bs in 3G, eNodeBs in 4G) to the core network. The backhaul

network connects the base stations to the mobile core, and MPLS is employed to provide efficient packet forwarding, QoS, and traffic engineering capabilities in this segment.

- Seamless MPLS: Seamless MPLS is a technology introduced to simplify MPLS deployments in mobile backhaul networks. It enables the seamless integration of MPLS with legacy time-division multiplexing (TDM) and Ethernet-based technologies used in the access networks. This allows for a smooth migration from older technologies to an MPLS-based transport infrastructure while maintaining backward compatibility.
- Hierarchical MPLS (H-MPLS): Hierarchical MPLS is an architecture that provides hierarchical LSPs in the transport network. It enables efficient scaling of MPLS networks by allowing multiple levels of MPLS domains and LSPs. H-MPLS is often used in 4G networks to support the aggregation of base stations and the efficient transport of traffic towards the core network.

Just as mobile services and technologies have evolved from circuit switched voice services to packet-technology-based data and voice services, the transport network has evolved rapidly from a purely time division multiplexing (TDM) infrastructure to a packet-based infrastructure.

The unified multiprotocol label switching (U-MPLS) solution is designed around standards-based packet technologies and carried over a high-capacity optical infrastructure. Globally, mobile operators have comprehensively adopted unified MPLS, setting the standard for IP-based mobile service delivery.

There are three fundamental aims of unified MPLS. The first is obvious: to build an infrastructure able to support the packet and TDM services needed for a 2G/3G/4G mobile network. The second is scalability, such that the architecture can support the networking requirements of the very largest operators using packet technology starting from the core all the way to the access layer of the network. The third is building a cost-effective infrastructure, so that as you move from the core toward the access, the forwarding and control plane of the network equipment can become simpler and more cost-effective.

Unified MPLS relies on splitting the network into different domains and introducing a hierarchy that has an IP core domain (not to be confused with the mobile core) in the center, with aggregation and access domains surrounding it. Each domain is isolated from the others and runs its own interior gateway protocol (IGP). Where inter-domain connectivity is required to deliver a service, border gateway protocol (BGP) carries the appropriate inter-domain prefixes, allowing communication from end to end.

This technique has proven to be extremely successful and many 3G and 4G networks are deployed today using unified MPLS running in the core, metro, and access networks. Currently, some mobile networks that use this architecture support more than 150,000 network devices.

5.3 Evolution of IP Transport to Support 5G Services

Although extremely successful and widely deployed, there are complexities to the unified MPLS architecture when deployed in extremely large networks. These revolve around the complexity of the control plane, the number of control plane protocols, and the number of device-level configurations required when building a service that spans many domains.

The introduction of 5G introduces additional challenges, which include:

- Footprint expansion due to the densification of the radio.
- Need to incorporate the network or edge data center seamlessly into the transport network.
- Use of virtualized network functions (VNF) for radio and mobile core functions.
- Network slicing.
- Control plane and user plane separation (CUPS) introduced by the CUPS architecture.

These changes, plus a desire by many operators to build a single converged transport architecture for their entire customer base, reinforces the need for an end-to-end packet infrastructure that can support a wide range of layer 2 and layer 3 services.

The proposed solution is to evolve the transport network's control and data plane toward segment routing, with the option of moving toward an infrastructure orientated around software-defined networks (SDN). This approach reduces control plane protocols, such as the label distribution protocol (LDP) and the resource reservation protocol – traffic engineering (RSVP-TE), removes flow state from the network, and provides several options for control plane implementation. These options range from a fully distributed implementation to a hybrid approach where the router and the SDN controller divide the functionality between themselves.

This transport infrastructure can overlay a wide range of service technologies, with associated IP-based service-level agreements (SLAs), including BGP-based VPN technologies, such as EVPNs and VPNv4/v6s, and emerging SD-WAN VPN technologies.

The 5G xHaul transport network refers to the infrastructure and technologies used to transport data and information in a 5G network. x-Haul, short for "any-haul" or "cross-haul," encompasses the various transport technologies used to connect the different components of a 5G network, including radio access network (RAN) nodes, core network elements, and data centers.

In a 5G network, xHaul provides high-speed and low-latency connectivity between the base stations and the core network, enabling the delivery of the advanced capabilities and services that 5G promises. It plays a crucial role in ensuring the seamless transmission of massive amounts of data, ultra-low latency applications, and supporting the diverse requirements of different 5G use cases, such as enhanced mobile broadband (eMBB), massive machine-type communications (mMTC), and ultra-reliable low-latency communications (URLLC).

The key characteristics and requirements of a 5G xHaul transport network include:

- High capacity: 5G networks demand significantly higher data rates compared to previous generations. x-Haul networks need to provide sufficient capacity to handle the increased data traffic generated by 5G devices and applications.
- Low latency: 5G aims to reduce latency to a minimum typically 20–50 ms in the fronthaul network and around 100 ms in the backhaul network to support real-time applications such as autonomous vehicles, remote surgery, and augmented reality. xHaul transport networks should minimize latency to ensure fast and responsive communication between different network elements.
- Flexibility and scalability: 5G networks are designed to support a wide range of services and applications. The xHaul transport network should be flexible and scalable, allowing for the addition of new network elements and the adaptation to changing traffic patterns and capacity requirements.
- Reliable and resilient: 5G is expected to provide high reliability and availability. The xHaul transport network should be designed with redundancy and resilience mechanisms to ensure continuous connectivity even in the presence of failures or network congestion.
- Convergence: 5G networks aim to converge different types of traffic, such as mobile broadband, IoT, and mission-critical applications, onto a single network infrastructure. The xHaul transport network needs to support this convergence by efficiently handling diverse traffic types and quality-of-service requirements.

To meet the requirements and characteristics of 5G network and connectivity, the 5G xHaul transport network utilizes various technologies and approaches (Figure 5.1), such as:

- Fiber optics: Fiber-optic cables are used to provide high-bandwidth and low-latency connections between the base stations and the core network. Fiber-optic infrastructure offers the capacity and speed required to handle the massive data traffic generated by 5G networks.
- Ethernet: Ethernet-based transport technologies, such as Carrier Ethernet, are used to provide scalable and flexible connectivity for 5G xHaul. Ethernet enables efficient data transport and supports the stringent requirements of 5G, including low latency and high reliability.

Figure 5.1: 5G transport overview.

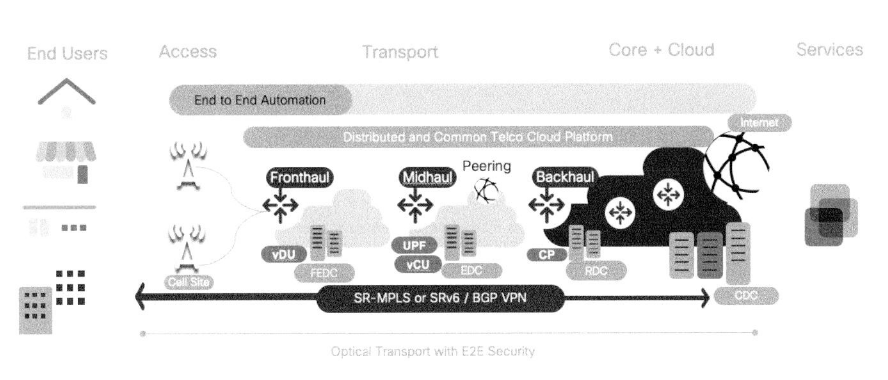

- Microwave and millimeter wave (mmWave): In certain scenarios where fiber deployment is challenging or uneconomical, wireless technologies like microwave and mmWave can be used for xHaul connectivity. These wireless solutions provide high-speed connectivity over the air, enabling rapid deployment and flexibility.
- Edge computing: Edge computing is an essential component of the 5G xHaul transport network. By placing computing resources closer to the edge of the network, near the base stations or cell towers, latency can be significantly reduced. This enables applications that require real-time processing and immediate response times.
- Network slicing: Network slicing allows the partitioning of a single physical network into multiple virtual networks, each tailored to specific 5G use cases. xHaul networks can be sliced to allocate dedicated resources and quality of service (QoS) parameters for different applications and services.
- Software-defined networking (SDN) and network function virtualization (NFV): SDN and NFV technologies play a crucial role in the 5G xHaul transport network by enabling dynamic management, orchestration, and virtualization of network functions. These technologies enhance network flexibility, scalability, and agility, making it easier to adapt to changing requirements and traffic patterns.
- Synchronization: In a 5G network, synchronization refers to the process of aligning the timing and frequency of various network elements to ensure proper operation and coordination. Synchronization is crucial for efficient and reliable communication within the network.

Overall, the 5G xHaul transport network forms the backbone of 5G connectivity, enabling the efficient and reliable transmission of data between the base stations and the core network. Its high-capacity, low-latency, and flexible characteristics are essential for delivering the full potential of 5G services and applications.

Figure 5.2: Transport evolution.

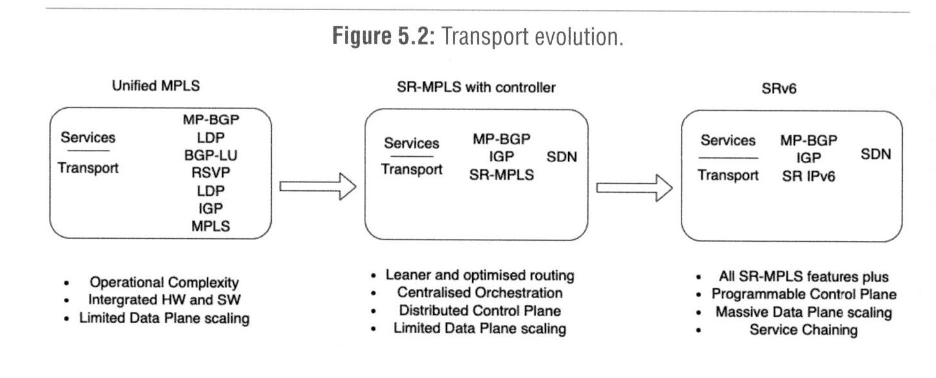

Figure 5.2 explains the evolution of transport fabric over time to cater for the various needs of end user traffic like data, voice and video. Unified MPLS transport which caters mostly to 4G services has operational complexity managing multiple protocols stack, no centralized path controller, limited convergence, network slicing for differentiated service treatments, and limited data plane scaling.

5.4 5G Transport Use Cases

The converged, end-to-end packet infrastructure, beginning in the access layer and stretching via the network data center all the way to the core, based upon segment routing and packet based QoS, provides the underlying xHaul transport network. This provides the most flexibility of application placement, the best scalability, the most robust reliability, and the leanest operational costs. On top of this, we layer VPN services, either based on BGP-based VPNs or business software-defined WAN (SD-WAN) technologies, to provide the means to support a multi-service environment capable of supporting strict SLAs.

This architecture provides transport capabilities needed to support the following use cases and capabilities:

- Converged, meaning it should be able to concurrently support:
  - Fixed and mobile traffic
  - Enterprise, small and medium businesses (SMB), IoT, and consumer services
  - Retail and wholesale business models
  - Voice, video, and data.

- End-to-end packet-based infrastructure based on an underlay of segment routing.
- Enables the encapsulation and transport of legacy RAN protocols (e.g. CPRI), and supports all possible fronthaul RAN, midhaul RAN, and backhaul architectures.
- Flexibility, supporting a range of services, including high bandwidth, best effort, and low-latency applications.
- High-bandwidth WAN network: optical infrastructure overlaid with high-performance hardware routers and switches that can deliver massive increases in throughput while retaining feature-efficient power and space footprints
- Virtualized service functions and user applications, deployed using commercial off-the-shelf (COTS) equipment in distributed, network-orientated data centers.
- Delivers to strict SLA requirements, controlling an end-to-end data forwarding plane that encompasses the WAN to the data center.
- VPN-capable, implementing network-based BGP layer 1/layer 2/layer 3 VPNs combined with support for emerging customer premises equipment (CPE) based SD-WAN solutions. In both cases, they should be able to utilize capabilities of the underlying network to meet SLA expectations based on QoS classes and traffic engineering.
- Accurately timed and synchronized in frequency, phase, and time, for all network topologies and transport types.
- Able to support open, multi-vendor RAN architectures such as O-RAN.
- Orchestrated and automated with model-driven provisioning and monitored with advanced analytics and model-based telemetry.
- Security embedded in the network, and built upon a chain of trust.

The next section explains how software defined networking (SDN) based transport with segment routing technology helps to achieve the 4G applications use cases and also unleash the potential of 5G services.

5.5 SDN Transport with Segment Routing to Support 5G Services

The change of transport technology to segment routing over MPLS data plane (SR-MPLS) where a massive protocol stack simplification happened with leaner and optimized routing, centralized SDN controller orchestration for end-to-end topology visibility and creating dynamic SLA based paths, 50 ms convergence with topology independent – loop free alternate (TI-LFA), and a distributed control plane.

In a 4G network, segment routing MPLS (SR-MPLS) can be used for transport to provide efficient and scalable packet forwarding. SR-MPLS combines the benefits of MPLS and segment routing, allowing for explicit path control and traffic engineering.

Figure 5.3: Transport overview.

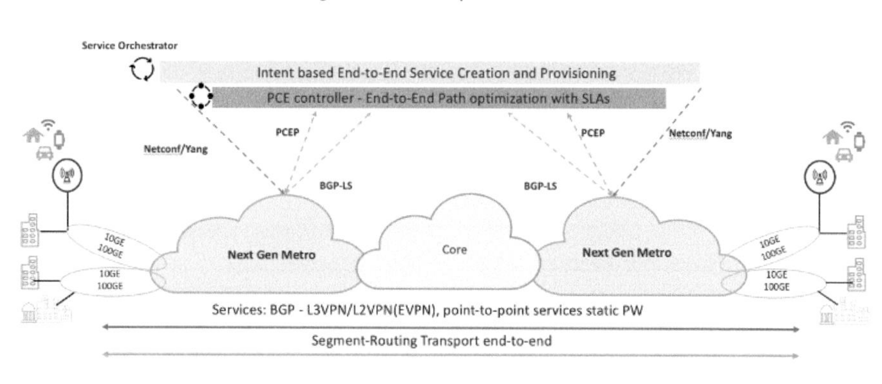

Figure 5.3 gives an overview of SR-MPLS transport in a 4G network:

MPLS-based backhaul: In a 4G network, MPLS is commonly used for the backhaul transport of traffic from the base stations (eNodeBs) to the core network. MPLS offers efficient packet forwarding, quality of service (QoS) support, and scalability, which are crucial for transporting the high volumes of data generated by 4G networks.

Segment routing (SR): SR is a network architecture that simplifies the forwarding of packets by encoding routing instructions in the packet header itself. In SR, the path through the network is determined by a sequence of segments. Each segment represents a specific network node or function.

SR-MPLS encapsulation: In SR-MPLS, the data packets originating from the base stations are encapsulated with MPLS labels and SR segments. The MPLS labels are used for forwarding and switching within the MPLS network, while the SR segments define the desired path through the network.

Explicit path control: SR-MPLS allows for explicit path control, where the network operator can define and control the exact path that packets will take through the network. This control enables efficient traffic engineering, load balancing, and optimization of network resources for ultra-reliable low latency (uRLLC) 5G applications with data sensitive traffic.

Traffic engineering: With SR-MPLS, network operators have fine-grained control over traffic engineering. They can define specific paths based on policies, QoS requirements, or network conditions. Traffic can be directed through specific nodes, avoiding congested or underperforming links, to optimize network performance. This helps in logical separation of the network to cater specific 5G application needs with specific SLAs and differential treatment.

Fast re-route (FRR) with TI-LFA: SR-MPLS supports fast re-route mechanisms to ensure network resilience and minimize service disruptions in the event of link or node failures. By precomputing backup paths, SR-MPLS can quickly redirect traffic to an alternative path upon detecting a failure, reducing downtime. This is achieved by using topology independent – loop free avoidance (TI-LFA) technology where in any topology the pre-programmed back-up paths helps to achieve convergence within 50 ms for 5G use cases of uRLLC applications and also massive machine-to-machine type communications (mMMTC).

Seamless integration with MPLS: SR-MPLS can seamlessly integrate with traditional MPLS networks, allowing for the coexistence of SR and non-SR traffic. This integration enables a smooth migration from traditional MPLS-based networks to SR-MPLS, leveraging existing infrastructure and investments.

Network slicing using SR-IGP flex algo: Network slicing using the segment routing (SR) flexible algorithm provides a flexible and efficient approach for creating and managing virtual network slices. Here's how network slicing can be achieved using SR's Flexible Algorithm:

Slice identification: Each network slice is assigned a unique identifier, which is encoded as a segment in the SR header of packets belonging to that slice. This identifier distinguishes traffic belonging to different slices and enables the network to route packets accordingly.

Flexible algorithm configuration: The SR flexible algorithm is configured with different objective functions and constraints that define the requirements and characteristics of each network slice. These objective functions can be based on factors such as latency, bandwidth, QoS, or any other metric relevant to the slice's specific needs.

Slice-aware path computation: The SR flexible algorithm dynamically computes slice-aware paths for packets based on the configured objective functions and constraints. When a packet enters the network, the algorithm determines the optimal path for that packet, taking into account the specific requirements of the slice it belongs to. This ensures that traffic within each slice is routed along the desired path and receives the appropriate treatment.

Resource allocation and isolation: The SR flexible algorithm enables efficient resource allocation and isolation between network slices. Network operators can allocate specific resources, such as bandwidth, processing capacity, or network functions, to each slice based on its requirements. This ensures that each slice has dedicated resources and operates independently, providing isolation and meeting the performance needs of the applications or services running within the slice.

Dynamic adaptation and optimization: The SR flexible algorithm can dynamically adapt to changes in network conditions and traffic demands. It continuously monitors the state of the network, including link utilization, latency, and other metrics. Based on this real-time information, the algorithm can dynamically recompute the optimal paths for each slice, ensuring efficient resource utilization and optimal performance even in dynamic environments.

By leveraging the flexibility and programmability of SR's flexible algorithm, network slicing can be achieved with customized routing, resource allocation, and isolation for different slices. This approach allows for efficient management of virtual network slices, tailored to specific use cases, and enables operators to optimize network resources and provide differentiated services to meet the diverse requirements of applications and users.

In addition , these are some of the key benefits that segment routing brings and helps in 4G and 5G networks:

- Multi domain policy and slicing using a path computation controller.
- Per destination on-demand nexthop (ODN) with automated steering.
- Per flow automated steering.
- Liveness monitoring of SR policy.
- Per link delay and performance measurement.

SR-MPLS in 4G networks provides benefits such as explicit path control, efficient traffic engineering, and network resilience. It offers a scalable and flexible transport solution to meet the increasing demands of mobile traffic and enhance the performance of 4G networks.

Segment routing with MPLS is an ideal technology for a converged transport architecture, since it can address the requirements of 5G, while simultaneously supporting fixed, enterprise, and consumer services. It provides:

- An evolutionary approach when moving from classic MPLS to SR/MPLS.
- Protocol simplification, removing the need for RSVP-TE and LDP.
- Reduction of the amount of state held on the network routing equipment.
- Concurrent support for services with different SLA requirements.
- Minimal configuration due to functions such as path computation element (PCE), automatic steering, and ODN.
- High scalability, allowing support for very large multi-domain networks.
- Operator choice on the level of centralized controller (SDN) involved in provisioning and operating the network.
- A strong linkage between the service layer and the underlying segment routing transport network through coloring of service routes. In this instance, coloring is not a literal term and refers to the ability to automatically specify the SLA requirements of the service. In the case of BGP VPNs, this is achieved through the color extended community attribute as specified in RFC5512.

5.6 Evolution of Segment Routing to the IPv6 Data Plane which Further Simplifies the Capabilities of 5G Services

The SDN based transport simplification journey further evolved with segment routing unleashing the potential of the IPv6 data plane. This offer several benefits for 5G use cases with optimized performance for various 5G applications, contributing to the success of the next-generation network deployments enabling the efficient and reliable delivery of services and applications.

IPv6 is the next generation of IP addressing, and has been available for more than two decades. IPv6 promised simplified networks and services by utilizing the large amount of address space to use IPv6 addressing to easily correlate packet to service. The vision has been there, but the proper technology did not fully enable it. Segment routing over IPv6 or SRv6 is the technology which not only provides an IPv6-only data plane across the network, but also creates a symmetry between data plane, overlay services, performance monitoring and enabling next-generation IPv6 based networks to support complex user and infrastructure services.

In addition to the simplification benefits that SR-MPLS brings, these are the key value proposition of segment routing over IPv6 transport:

- Simplicity at the IP layer by merging underlay and overlay using SRv6 next C-SID.
- Transit nodes can do simple IPv6 forwarding without being SRv6 aware.
- Network as a program for 5G applications slicing.
- Service chaining.
- Inbuilt load balancing with flow-label in the IPv6 header.

- Prefix summarization with excellent scalability.

Leveraging the above-mentioned capabilities, SRv6 networks offer several benefits for 5G use cases, enabling the efficient and reliable delivery of services and applications. Here are three key advantages of SRv6 networks in the context of 5G:

Enhanced network slicing: Network slicing is a critical feature of 5G that allows the creation of virtualized networks tailored to specific use cases. SRv6 provides a flexible and scalable solution for network slicing by leveraging the capabilities of IPv6 and segment routing. It enables the creation of virtual topologies, allowing operators to allocate resources and define specific routing paths for each slice. This granular control over the network ensures efficient resource utilization, improved isolation between slices, and optimized performance for different 5G use cases.

Improved latency and QoS: 5G demands low-latency connectivity and stringent quality of service (QoS) requirements for applications like autonomous vehicles, virtual reality, and industrial automation. SRv6 transport offers optimized packet forwarding and traffic engineering capabilities, enabling low-latency routing and efficient resource allocation. By leveraging the source routing paradigm, SRv6 allows for deterministic forwarding paths, reducing packet processing time and optimizing latency-sensitive applications. Additionally, SRv6's ability to assign different service levels and QoS parameters to specific segments ensures that critical applications receive the required performance guarantees.

Scalability and flexibility: 5G networks are expected to support a massive number of connected devices and handle exponentially increasing data traffic. SRv6 brings scalability and flexibility to network deployments by leveraging the routing capabilities of IPv6 and segment routing. It simplifies network operations by reducing the complexity associated with traditional routing protocols, enabling efficient packet forwarding and optimized resource utilization. SRv6's flexibility allows network operators to adapt their infrastructure to changing traffic patterns and evolving service requirements, ensuring the scalability and future-proofing of 5G networks.

SRv6 networks provide enhanced network slicing capabilities, improved latency and QoS, as well as scalability and flexibility to meet the demands of 5G use cases. By leveraging the benefits of segment routing and IPv6, SRv6 enables efficient resource allocation, low-latency connectivity, and optimized performance for various 5G applications, contributing to the success of the next-generation network deployments.

5.7 Clocking in a 5G xHaul Transport Network

Clocking in a 5G xHaul transport network is crucial for maintaining synchronization and ensuring the proper functioning of the network components.

Here are some reasons why clocking is important for 5G services:

Synchronization of base stations: 5G networks rely on a dense deployment of base stations, including small cells and massive MIMO antennas. These base stations need to be synchronized to avoid interference and ensure efficient coordination among neighboring cells. Accurate clocking allows for precise synchronization, enabling coordinated transmission and reception across the network, which is critical for achieving high data rates, low latency, and reliable connectivity in 5G.

Time division duplex (TDD) operation: 5G networks often employ TDD, where the same frequency is used for both uplink and downlink transmissions, with the time division between them. To avoid interference and ensure efficient use of the available spectrum, accurate clocking is essential. It enables precise synchronization of the time slots for uplink and downlink transmissions, ensuring that they are correctly aligned and do not overlap, thereby maximizing spectral efficiency.

Network slicing: 5G introduces the concept of network slicing, where the network is divided into virtualized slices, each tailored to specific use cases or applications. Synchronization and accurate clocking are crucial for network slicing, as different slices may require different levels of quality of service (QoS) and timing requirements. Proper clocking ensures that each slice receives the required timing parameters, such as latency or synchronization accuracy, to meet the specific needs of the services or applications running within the slice.

Low latency applications: 5G enables various low latency applications, such as autonomous vehicles, real-time gaming, and industrial automation. Accurate clocking is essential for achieving low latency in these applications. Precise synchronization ensures that signals are transmitted and received at the right time, minimizing delays and providing real-time responsiveness. Clocking accuracy directly impacts the ability of 5G networks to deliver ultra-low latency, which is critical for supporting time-critical applications.

In summary, clocking is important for 5G services as it enables synchronization of base stations, supports TDD operation, facilitates network slicing, and ensures low latency for various applications. Accurate and synchronized timing is essential to maximize the performance, efficiency, and reliability of 5G networks, ultimately delivering an optimal user experience and enabling the full potential of 5G services.

Due to the inherent need for precise time synchronization 3GPP defined TSN (time sensitive networking) is the preferred architecture for the 5G system (5GS). The TSN is defined by a set of standards under IEEE802.1 sub-group. Further detail about TSN can be found at https://1.ieee802.org/tsn/ . The TSN specifies among other things, time synchronization for fronthaul, time sync for time sensitive applications, QoS and flow control, etc. The 5GS implementation can be achieved using a combination of boundary clock and edge grandmaster.

Each TSN reference clock shown in Figure 5.4 can be replaced with a cost effective boundary clock/edge grandmaster. The grandmaster provides TSN profiles for fronthaul synchronization and the same device can act as a boundary clock or a grandmaster (Figure 5.5). Understanding the concept of 5GS is important as it implies the need for highly distributed and precise time synchronization to keep 5G network optimized and operational.

Figure 5.4: Transport clocking overview.

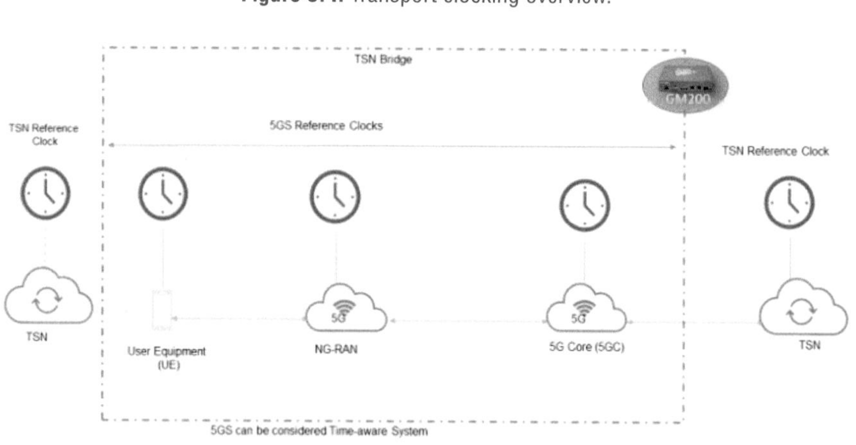

Synchronization is critical in 5G networks and more importantly in the fronthaul design. The O-RAN alliance has defined four types of S-plane (synchronization plane) configuration modes for timing distribution in the RAN infrastructure. The S-Plane configuration modes are specified in O-RAN control, user and synchronization plane specification (O-RAN.WG4.CUS.0-v05.00) and mainly addresses sync plane configuration between O-RU and O-DU. These configuration modes are known as follows:

Figure 5.5: Transport clocking flow.

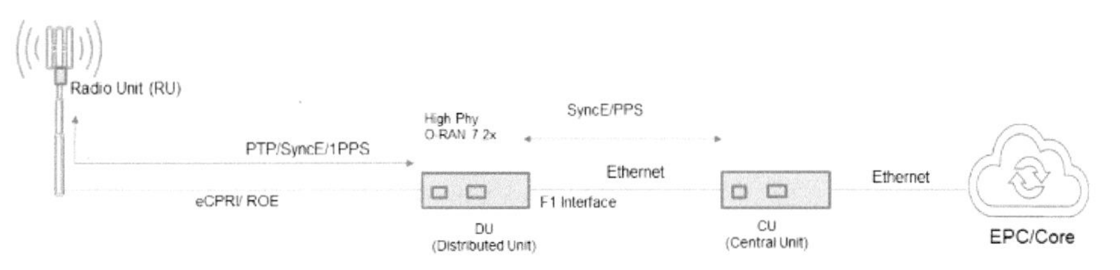

- **Configuration LLS-C1 (LLS-C1):** This configuration specifies network timing distribution from O-DU to O-RU via point-to-point topology between central site and remote site [1].
- **Configuration LLS-C2 (LLS-C2):** In this configuration, one or more ethernet switches are allowed for network timing distribution from O-DU to O-RU between central sites and remote sites. The interconnection among switches and fabric topology (for example mesh, ring, tree, spur etc.) are out of scope of this configuration and subject to deployment decisions [1].
- **Configuration LLS-C3 (LLS-C3):** In this setup, network timing distribution is done from PRTC/T-GM to O-RU between central sites and remote sites. One or more Ethernet switches are allowed in the fronthaul network. Interconnection among switches and fabric topology (for example mesh, ring, tree, spur etc.) are deployment decisions which are out of the scope of O-RAN specification [1].
- **Configuration LLS-C4 (LLS-C4):** In this configuration local PRTC (primary reference time clock) provides timing input to O-RU [1].

Two primary types of clocking are relevant in this context: phase synchronization and frequency synchronization.

Phase synchronization: Phase synchronization ensures that different network elements operate in phase with each other. This is especially important in 5G networks due to the use of advanced technologies like massive MIMO (multiple-input multiple-output) and beamforming, which require precise coordination between base stations.

To achieve phase synchronization in a 5G xHaul network, the precision time protocol (PTP), defined in the IEEE 1588 standard, is commonly used. PTP allows network devices to synchronize their clocks with high accuracy. In the context of 5G xHaul, the grandmaster clock (usually located at the core network or data center) provides timing information to the base stations and other network elements.

PTP (precision time protocol) clocking plays a crucial role in 5G xHaul networks, providing accurate and synchronized timing across the network.

Synchronization of remote radio heads (RRHs): In 5G xHaul networks, remote radio heads (RRHs) are deployed at the edge of the network to support high-speed wireless connectivity. These RRHs require precise synchronization to ensure coordinated transmission and reception of signals. PTP clocking enables accurate time synchronization between the RRHs and the central baseband unit (BBU), ensuring proper coordination and minimizing interference, ultimately improving the overall network performance.

Timing distribution for time-sensitive applications: 5G xHaul networks support various time-sensitive applications, such as industrial automation, smart grid, and real-time communication. These applications often require strict timing requirements and synchronization accuracy. PTP clocking provides the necessary timing distribution mechanism to ensure that time-sensitive applications receive accurate timing information, minimizing latency and enabling reliable and predictable performance.

Coordinated operation in multivendor environments: 5G xHaul networks are often composed of equipment from different vendors. Each vendor's equipment may have its own clock source and timing capabilities. PTP clocking provides a standardized protocol for clock synchronization, allowing different vendor equipment to synchronize their clocks accurately and maintain proper coordination. This ensures seamless interoperability and coordinated operation across the network, regardless of the equipment vendor.

Frequency synchronization: In addition to time synchronization, PTP clocking also facilitates frequency synchronization in 5G xHaul networks. Frequency synchronization is important for carrier aggregation, where multiple frequency bands are combined to achieve higher data rates. PTP enables the alignment of the carrier frequencies across different network nodes, ensuring proper frequency coordination and efficient spectrum utilization.

Network expansion and scalability: PTP clocking supports network expansion and scalability in 5G xHaul networks. As the network grows and new network elements are added, PTP enables the seamless integration and synchronization of these new elements into the existing network. It allows for the dynamic discovery and management of clock sources, simplifying the deployment and maintenance of the network as it evolves.

PTP clocking is crucial in 5G xHaul networks for synchronization of RRHs, timing distribution for time-sensitive applications, coordinated operation in multivendor environments, frequency synchronization, and network expansion. Accurate and synchronized timing provided by PTP clocking enhances the performance, reliability, and scalability of 5G xHaul networks, enabling the delivery of high-quality services and applications.

Frequency synchronization: Frequency synchronization ensures that network elements share a common frequency reference. This is crucial for maintaining the coherence of transmitted signals and avoiding interference between neighboring cells.

The synchronous Ethernet (SyncE) standard is commonly used for frequency synchronization in 5G xHaul networks. SyncE utilizes Ethernet physical layer attributes to distribute a common frequency reference across the network.

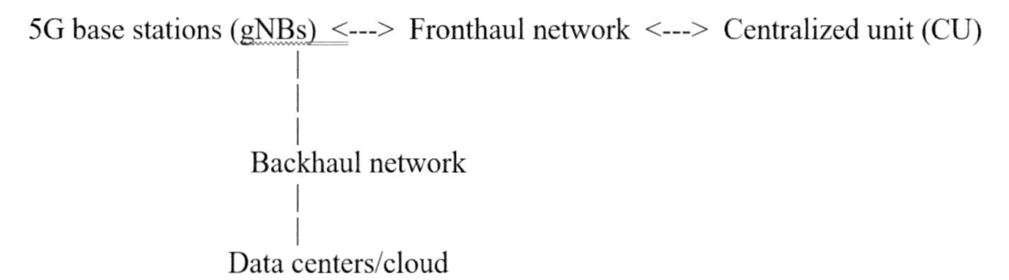

SyncE clocking works in 5G xHaul networks by distributing a highly accurate and stable clock signal throughout the network to ensure precise frequency and phase synchronization. Here's a simplified overview of how SyncE clocking operates in a 5G xHaul network:

Master clock source: A master clock source is selected in the network, typically a highly accurate and stable clock, such as an atomic clock or a GPS receiver. The master clock serves as the primary reference for synchronization in the network.

Clock generation: The master clock source generates a high-quality clock signal that adheres to the required synchronization accuracy. This clock signal is then distributed to the SyncE-enabled network elements, including network switches, routers, and other network devices.

SyncE interfaces: SyncE interfaces are used to transmit and receive the clock signal within the network. These interfaces can be physical Ethernet interfaces or dedicated synchronization interfaces that support the SyncE protocol.

Clock distribution: The master clock signal is distributed to SyncE-enabled devices over dedicated synchronization links or embedded within the Ethernet data traffic. SyncE utilizes the Ethernet physical layer to distribute the clock signal in a synchronous manner.

Slave clocks: SyncE-enabled devices act as slave clocks and receive the distributed clock signal from the master clock. They use this received clock signal as a reference to synchronize their local clocks.

Clock recovery and compensation: The slave clocks recover the clock signal and perform compensation to align their local clocks with the received clock signal. SyncE employs mechanisms such as phase-locked loops (PLLs) and digital signal processing techniques to ensure accurate synchronization and compensate for any clock drift or variation.

Synchronization monitoring: SyncE-enabled devices continuously monitor the received clock signal and compare it with their local clock. They measure the timing accuracy and adjust their clock synchronization accordingly to maintain precise frequency and phase alignment.

Redundancy and resilience: SyncE supports redundancy mechanisms to ensure high availability and resilience in case of clock source failures or network disruptions. Redundant clock sources and distribution paths, along with synchronization supply units (SSUs) and synchronous Ethernet equipment clocks (EECs), help maintain synchronization during failures and provide backup clock sources.

By following these steps, SyncE clocking ensures accurate and synchronized timing across the 5G xHaul network, enabling efficient transmission and reception of signals, coordination between network elements, and reliable operation of time-sensitive applications.

The clocking mechanisms would be integrated into the fronthaul and backhaul networks to ensure accurate synchronization throughout the network. Each network element would receive timing information from a central source (usually a grandmaster clock) to maintain synchronization. The synchronization methods mentioned, PTP for phase synchronization and SyncE for frequency synchronization, help achieve the required accuracy for time-sensitive applications like 5G.

Proper clocking mechanisms are essential to ensure that the 5G xHaul network can support the low-latency and high-capacity requirements of 5G services, such as ultra-reliable communication and massive IoT connectivity.

CHAPTER

6

Telco Cloud

6.1 Telco Cloud – What is it and How is it Different from an IT Cloud?

Cloud computing refers to the provision of computing services via the internet, encompassing elements such as servers, storage, databases, networking, software, analytics, and intelligence. These services are commonly referred to as "the cloud" and are aimed at facilitating quicker innovation, adaptable resource allocation, and cost-efficient scalability. Instead of owning and maintaining their own physical servers and data centers, users can rent or subscribe to these resources from cloud service providers.

In the service provider world, "Telco Cloud" is used to represent a private cloud deployment within the Telco environment that hosts virtual network functions (VNFs) or recently CNFs (cloud-native network functions) of the Telco Network utilizing NFV/NFC techniques.

6.1.1 Challenges that triggered Telco Cloud development

Telcos have been deploying their network (core/RAN) using monolithic architectures for many decades. In this approach, network services and functions are tightly integrated into a single, often proprietary, hardware and software stack. This created multiple challenges for Telcos to evolve at the pace of growth or change the industry was going through.

Exponential growth in traffic: Usage pattern has evolved from simple mail/browsing to heavy video, audio and gaming in the past decade. These use cases consume high bandwidth. Adding the large number of IOT devices to this, the traffic in the current Telco network is huge compared to what has been some years ago.

Scalability: Telco services often face fluctuations in demand, such as peak usage times and unexpected traffic spikes. With monolithic architecture, it is not possible to scale at the pace required.

Cost efficiency: Capital and operational expenditure is high with purpose-built hardware and software. Hardware reuse, resource optimization and re-alignment, etc. is not possible at ease in traditional deployments.

Service agility: The current dynamic market introduces new services and features rapidly with constantly evolving customer expectations. Traditional networks do not have the agility to allow quick deployment of these new use cases.

Low margins: Lower margins due to market competition and the fact that Telcos are becoming data pipes for OTT players have pushed the operators to find new ways to reduce costs, such as bringing the data source close to the delivery (edge). This cannot be achieved effectively at a scalable fashion with traditional networks.

Vendor lock-in: Monolithic architectures always pose the challenge of vendor lock-in. This causes delay in acquiring hardware/software or deploying new use cases, hampering the Telcos wish to grow at the pace required by the market. The multi-vendor approach in traditional networks could mean higher operational costs as staff need to be trained in each vendor's proprietary hardware/software/operation solutions.

All the above-mentioned challenges can be effectively managed by Telco cloud. From dynamically scaling up at a short notice to hosting the network functions from different vendors on the same cloud, the options that Telco Cloud opens to the operators are huge.

6.1.2 Telco Cloud structure

Hypervisors play a pivotal role within Telco Cloud. These software or firmware components abstract physical hardware, facilitating the creation of virtual machines (VMs) that can run multiple operating systems on a single physical server. There are two main types of hypervisors: Type 1, which runs directly on the bare-metal hardware, and Type 2, which runs on top of a host operating system (Figure 6.1).

Figure 6.1: Hypervisor types in virtualization.

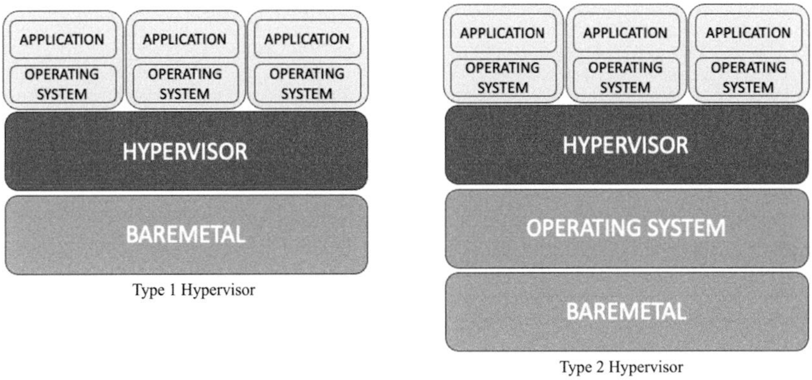

Type 1 Hypervisor

Type 2 Hypervisor

Type 1 hypervisors are commonly employed within Telco Cloud environments, providing a secure and high-performance foundation for virtualization. Hypervisors bring several benefits, including efficient resource utilization, hardware abstraction, isolation, snapshot and cloning capabilities, and support for multiple operating systems.

In the Telco Cloud landscape, virtual network functions (VNFs) are the building blocks of modern network services. These VNFs represent software-based counterparts to traditional hardware network functions, such as packet core network elements, firewalls, routers, and load balancers. By running as software on virtualized infrastructure, VNFs offer the flexibility to scale, orchestrate, and adapt services on the fly. They can be quickly instantiated, scaled, and chained together to create complex network service scenarios. For example, in a VNF-based network, data packets may traverse multiple VNFs in a specific order to provide security, optimization, and other services, all while being managed centrally and orchestrated through software.

Figure 6.2 shows the Telco Cloud architecture based on the ETSI NFV MANO stack. On the left side is the actual stack with hardware for compute, storage and networking with the hypervisor layer on the top projecting the virtualized resources towards the virtual network functions (VNFs). On the right side is the framework for managing and orchestrating virtualized network

Figure 6.2: Telco Cloud architecture based on ETSI NFV MANO stack.

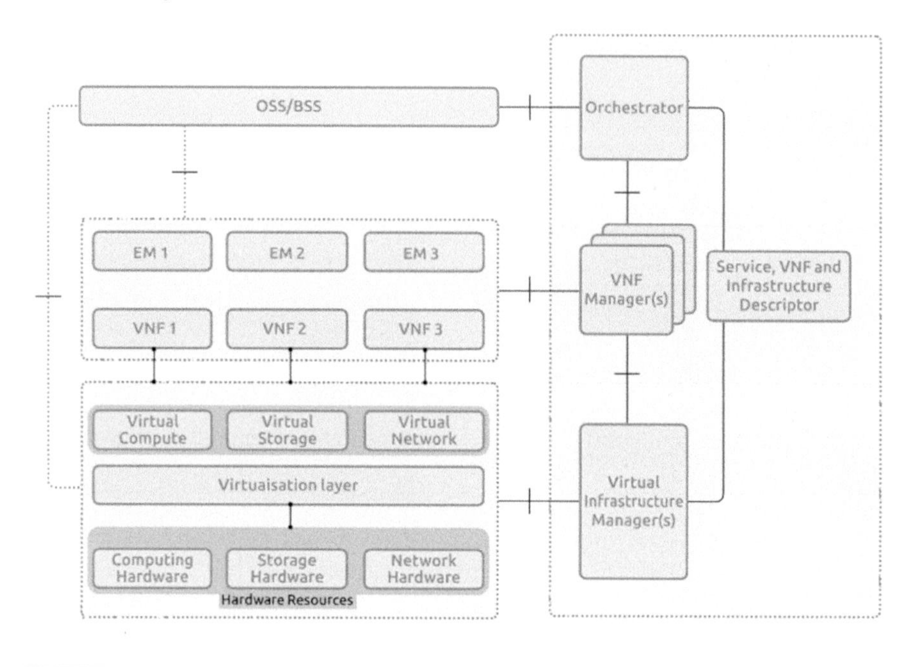

functions in a Telco Cloud environment. The NFV MANO stack consists of three key components:

NFV orchestrator (NFVO): The NFV orchestrator is responsible for coordinating and managing the lifecycle of VNFs and their resources. It interacts with the virtualized infrastructure manager (VIM) to allocate and manage the necessary compute, storage, and networking resources for VNFs. It communicates with the VNF manager (VNFM) to ensure proper instantiation, scaling, and termination of VNFs as required.

Virtualized infrastructure manager (VIM): The virtualized infrastructure manager is responsible for managing the underlying compute, storage, and network resources in the Telco Cloud infrastructure. It abstracts and virtualizes physical resources and provides a pool of resources that can be allocated to VNFs. It communicates with the NFV orchestrator to receive resource allocation requests and ensures the efficient utilization of the infrastructure.

VNF manager (VNFM): The VNF manager is responsible for managing individual VNFs throughout their lifecycle. It interacts with VNFs to ensure proper instantiation, scaling, and termination. It communicates with the NFV orchestrator to provide information about the status of VNF instances and to request the allocation or release of resources. The VNFM also manages fault and performance monitoring of VNFs.

These three components, NFV orchestrator, virtualized infrastructure manager, and VNF manager, work together to enable the dynamic provisioning, scaling, and management of virtualized network functions within the Telco Cloud environment. They are essential for ensuring the efficient and agile operation of virtualized network services in the telecommunications industry.

6.1.3 Differences between a Telco cloud and an IT cloud

The Telco Cloud is typically a private cloud setup within the Telco network to host the Telco workloads. In contrast, an IT Cloud primarily pertains to enterprise workloads and could run on mix of public, private, and hybrid clouds. An IT Cloud is designed to deliver cloud-based services that cater to the software needs of enterprises.

The Telco Cloud tries to ensure a secure and predictable traffic flow and has stringent latency requirements. Some Telco applications require extremely low delays (in single digit milliseconds), which is crucial for delivering on the promises of new 5G services. In contrast, while an IT Cloud also values low-latency performance, its latency requirements are generally not as strict as those of the Telco Cloud. Additionally, an IT Cloud often uses internet-based access, while Telco Clouds tend to rely on dedicated service provider links for delivering services over cloud.

Telco Cloud environments are based on open standards which allows for easy interworking between different vendor equipment. In contrast, IT and Enterprise Clouds mostly incorporate vendor-specific technologies.

For virtualization, the Telco Cloud often uses dedicated CPU allocation, whereas IT Clouds can distribute CPU resources among multiple applications.

6.2 Cloud Transition: From VNFs to CNFs

6.2.1 What is wrong with VNFs?

VNFs were not made for the cloud. The telecommunications vendors chose a “lift and shift” approach, which involved replicating existing embedded software systems designed for hardware functions and transferring them entirely into a single, occasionally cumbersome virtual machine (VM). Although this approach offered the advantage of cost savings by enabling the transition from dedicated proprietary hardware to more cost-effective commodity hardware, the migration to the cloud often did not result in the optimization of these virtual network functions (VNFs). Consequently, the VMs generated frequently remained inefficient, designed for specific tasks, and posed difficulties in terms of maintenance.

Because of the challenges, complexity, and lack of understanding at the Telco operator side, vendors have tried to replicate the "lock-in" by trying to sell "full stack" solutions. Although easier to deploy, these are still siloed and does not provide any of the advantages of the "openness" envisaged by the original concepts of the NFV and Cloud. Wherever the openness existed (due to certain strong-willed Telcos), it was found that the deployments were further complex due to the fact that most VNF vendors struggled to on-board their products, delivered poor performance and consumed excessive hardware resources when deployed.

Automating the deployment of VNFs is possible, but it's uncommon for VNFs to facilitate automated network address assignment. Consequently, a laborious manual process may be necessary for assigning IP addresses. Moreover, VNFs usually lack built-in service discovery mechanisms. As a result, when various components within a VNF need to communicate, configuring the necessary IP addresses for communication is required on each instance individually.

From a configuration management viewpoint, VNFs might have to perform operations on individual VM instances via a command line interface. Telco engineers must perform these manually, by typing commands which is error prone and tedious.

VNFs are mostly monolithic software architectures which support procedures like healing and scaling for namesake only. Since they have been ported from their hardware counterparts, there is a good amount of manual involvement in the procedures. Restoring a failed VNF could be complicated and requires human interaction. Similarly, upgrading a VNF is also a tedious process. Every new release could introduce a significant number of new features, and this brings the potential of disrupting the stability of the VNF service. As a result, a thorough and extensive testing phase is essential before implementing the release. Also, upgrading VNF software typically involves intricate procedures that must be executed one at a time, involving multiple command line-driven steps which are not easy to automate.

6.2.2 Function first (vertical) vs. platform first (horizontal) approaches

Many Telcos went for the function first approach for Telco cloud deployment. Because of the huge demand to increase the 4G capacity, and lack of maturity in the virtualization technology, they opted for the vertical "full stack" offerings from the vendors. With solutions coming in from existing vendors, Telcos found it easy to move to virtualization in the short term. Most software and interfaces remained the same as the existing purpose-built solutions, thus enabling Telcos to integrate it to their network quite easily. On the flip side, this still retained the dependency on a single vendor, even leading to multiple independent cloud deployments to support multiple vendors.

The platform first approach promotes first building a multi-vendor platform that allows horizontally integrated VNFs. This allows VNFs from different vendors to run alongside each other and tap the full potential of virtualization. It is not very simple as the technology is relatively new and interoperability issues are plenty. Not every vendor VNF is made to work with every cloud platform. Co-operation between vendors (driven by Telcos) will be required to achieve this. In addition, a lot of resources would be required for brainstorming, designing and building such a platform. Only a handful of operators who has the financial and technical resources can get this done overcoming these unfaced challenges.

6.2.3 Growing cloud with CNFs

In the previous sections we have seen how Telco Cloud has started off and the issues with virtualization as the choice of technology for the cloud. In this section we go into the details of containerization or "cloud-native" as the solution for the Telco Cloud. CNFs stands for cloud-native network functions. CNFs provide the same services and performs similar functions as their VNF counterparts – although in a cloud appropriate way. A CNF is a dynamically scalable, redundant system based on a collection of loosely coupled microservices. They are packaged in containers and ideally deployable without modification on any standard containers-as-a-service infrastructure (Figure 6.3).

CNFs are orchestrated by Kubernetes and use key components of the cloud-native ecosystem such as Helm charts, Prometheus, CNI, etc. Configuration management based on ConfigMap makes the solution easy to modify. In order to create a CNF that can efficiently exploit all the advantages offered by cloud-nativeness, it needs to be built from the ground up. Transforming a VNF, which

Figure 6.3: Container deployment stack.

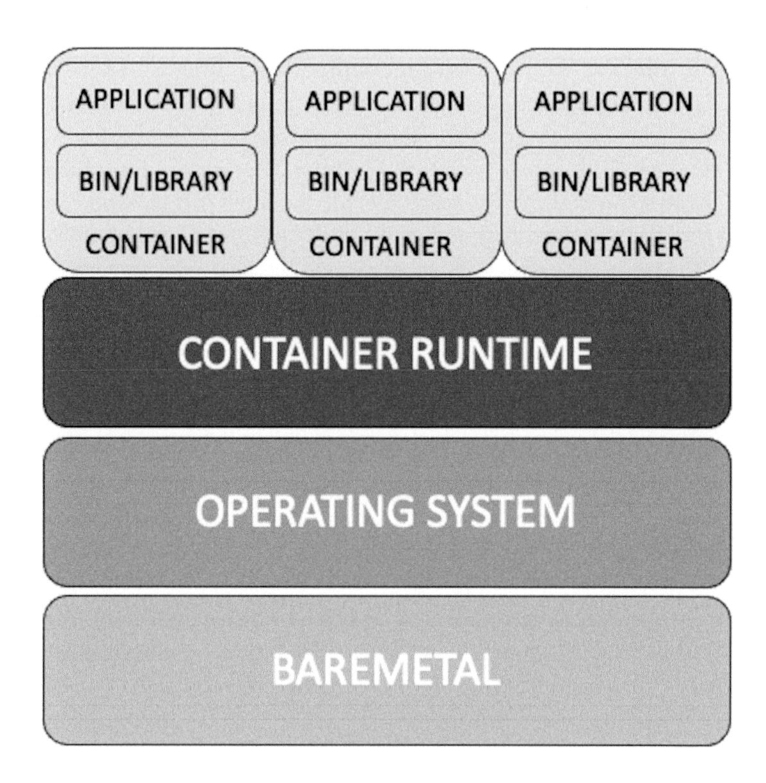

originally used software from a purpose-built machine, and placing it within a container, doesn't turn it into a CNF. Similarly, merely incorporating HTTP-based APIs into a VNF to enable a service-oriented architecture doesn't qualify it as a CNF. Let us take a look at the basic principles and characteristics of cloud-native NFs.

Containerization: CNFs are typically packaged in containers (e.g., Docker containers) to ensure consistency and portability. Containers make it easier to deploy, scale, and manage network functions across different cloud environments.

Microservices architecture: CNFs are often designed as microservices, breaking down complex network functions into smaller, more manageable components. Each microservice performs a specific task, which makes it easier to develop, maintain, and scale the network function.

Orchestration: CNFs are orchestrated using tools like Kubernetes. Orchestration allows for automatic scaling, load balancing, and high availability, making it easier to manage CNFs in a dynamic environment.

Automation: CNFs are built with a strong emphasis on automation. This includes automated scaling, configuration, and recovery. Automation reduces the need for manual intervention, making network management more efficient.

Stateless: CNFs are often designed to be stateless, meaning they don't store data about the state of the network or the applications they serve. Statelessness simplifies scaling and failover processes.

API-driven: CNFs typically provide well-defined APIs (application programming interfaces) that allow for easy integration with other network functions and applications. These APIs are HTTP-based and adhere to RESTful principles.

Service mesh: CNFs often leverage service mesh technologies like Istio or Envoy to manage communication, security, and monitoring between microservices.

Continuous deployment: CNFs embrace continuous deployment and continuous integration (CI/CD) practices, allowing for rapid updates and improvements to network functions.

Resilience: CNFs are designed to be resilient, with built-in redundancy and failover mechanisms to ensure high availability and minimal service disruption.

Cloud-native principles: CNFs follow cloud-native principles, such as DevOps, infrastructure as code (IaaC), and using cloud services when applicable. This helps in achieving the agility and scalability in the Telco Cloud.

Scalability: CNFs can easily scale horizontally to accommodate changing workloads and network traffic demands.

Elasticity: CNFs can automatically adjust their resource consumption based on demand, ensuring efficient resource utilization.

Monitoring and logging: CNFs include robust monitoring and logging capabilities, allowing operators to gain insights into the network function's performance and troubleshoot issues.

Service chaining: CNFs make it easier to implement service chaining, where network services are sequenced to process data traffic in a specific order.

This is valuable for applications like content filtering, security, and traffic optimization.

Edge computing: CNFs are well-suited for edge computing scenarios, where Telco services need to be deployed closer to the network edge for reduced latency and improved performance. This is particularly relevant in the context of 5G networks and IoT applications.

Cloud-native network functions represent a fundamental shift from VNFs in how network services are developed, deployed, and managed. They leverage containerization, microservices, automation, and other cloud-native principles to provide more flexible and efficient network solutions in the Telco Cloud.

6.3 Telco Cloud Architecture Considerations

6.3.1 Strategy and considerations on moving to Telco Cloud

Existing Telcos face a multitude of challenges to move to the cloud. The mindset of "If it's working, do not touch it" is a deterrent to this transformation. Telcos should be ready to experiment and try out combinations and permutations of options to host their workloads in a mix of VNF/Public/Private cloud environments. Telco's own digital environment has become complex over the years, leading to operational difficulties and inefficiencies that hampers seamless migration to cloud. Security is another aspect of worry for a Telco. Coming from a closed and secure environment to host the workloads, Telcos need good level of conviction and confidence to embrace public clouds to run their applications. Telcos will feel the lack of control over their own network they are habituated to have. Finally, having qualified experts (be it in planning or operations) is a pre-requisite to create the ideal transformation to cloud. However, lack of skill set in the newer technologies, competition from other Telcos, etc. pose a challenge to find appropriate human resources to develop, deploy and run the Telco cloud.

Whereas greenfield operators can build a completely cloud-native environment from the ground up, existing operators need to tune their migration strategy to have legacy and cloud-native networks coexist for some period, allowing them to migrate network functions, services and applications in a way that makes the most sense for their organization. The migration can be carried out on a step-by-step basis, focusing on individual network functions or services. However, it is crucial to initiate the process by conducting a thorough assessment of cloud readiness. This assessment should cover infrastructure, applications, service portfolios, organizational aspects and processes.

Telcos can choose to develop everything internally or build with a partner or even with a system-integrator who co-ordinates different partners and helps in technology selection. Depending on the type of workloads, timelines, budget, etc. they can adapt the approach.

Telcos should perform a cloud readiness assessment. They must define their VNF and application migration strategies. They need to demarcate workloads which should remain as VMs and those which can be developed into cloud-native microservices. Telcos should define the tools they can use to orchestrate and manage the cloud environment and these tools should support automation. A very important strategy is to determine the split of the applications that can be run on private and public clouds, based on parameters like economics, operational capacity and capability, architecture, ROI and time-to-market requirements. Finally, the ecosystem partner environment is very important in the cloud transformation. The willingness of the application partners to collaborate in the cloud journey providing high quality products, services, and support will ensure a successful transformation into the cloud for the Telcos.

6.4 Security in the Telco Cloud

6.4.1 General security risks and aspects to consider while deploying core network workloads in the Telco cloud

Security has become one of the most important factors that Telcos consider while choosing any new technology. Some of the challenges faced by them while adopting cloud are mentioned are as follows:

Data security and privacy: Telco Cloud environments host vast amounts of sensitive customer data, including call records, text messages, and personal information. Data breaches, data leakage, and privacy violations are major concerns.

Network security: The distributed nature of Telco Cloud networks and the interconnection of various components create an expanded attack surface. Protecting network infrastructure, routers, switches, and virtualized network functions is challenging, as any compromise can have heavy consequences.

Virtualization and hypervisor security: Virtualization technology is at the core of Telco Cloud environments. Securing hypervisors and ensuring the isolation of virtualized resources are essential, as vulnerabilities at this level can lead to catastrophic breaches. The same applies to cloud-native network functions also.

API security: APIs play a crucial role in Telco Cloud services, facilitating communication between various network functions. Inadequately secured APIs can be exploited, leading to unauthorized access and data leakage. Secure API development and management are essential.

Multi-tenancy: Many Telco Cloud environments host multiple tenants, including different service providers or enterprise customers. Ensuring the isolation and security of tenant resources is a complex challenge, as breaches by one tenant can potentially affect others.

Security of third-party components: Telco Cloud environments often rely on third-party components and services. Ensuring the security of these components, including software-defined networking (SDN) and network functions (NFV), can be challenging.

Orchestration and automation security: The orchestration and automation of network services can introduce new vulnerabilities. Unauthorized changes to network configurations or automation scripts can disrupt services.

To overcome these challenges, Telco service providers must approach security differently. They can start by considering the following strategy.

Micro-segmentation: Micro-segmentation is a cornerstone of Telco Cloud security. This strategy involves dividing the network into smaller, isolated segments, or microsegments, each with its own security policies and access controls. By doing so, operators can contain security breaches and prevent lateral movement within the network. This granular approach to security enhances overall network protection, particularly in a dynamic cloud environment.

Secure APIs: Application programming interfaces (APIs) are the lifeblood of Telco services, enabling seamless interaction between various systems. Ensuring the security of APIs is critical. This involves robust authentication, authorization mechanisms, and encryption. Secure APIs should adhere to best practices, like the use of API tokens and limiting access to authorized entities. Regular assessments and audits of API security are essential to identify and mitigate vulnerabilities.

Zero trust: The zero trust security model is gaining prominence in Telco Cloud security. Zero trust assumes that no entity, whether inside or outside the network, can be trusted by default. Access is based on the principle of least privilege and is continuously monitored. This approach helps safeguard against both insider and external threats, thereby enhancing network security.

Vulnerability management: Continuous vulnerability management is a proactive strategy that seeks to identify and address vulnerabilities in the Telco Cloud environment promptly. Operators should regularly scan for vulnerabilities, prioritize them based on their potential impact, and implement patches and remediation measures swiftly. This practice minimizes the window of opportunity for attackers and strengthens overall security.

Decouple access: Decoupling access involves separating user or device identities from the network or application they are accessing. Access can be controlled based on a user's identity, rather than their physical location or device. This approach allows for more flexible and secure access controls, particularly in a cloud environment where users may access services from various locations and devices.

As Telco operators continue to embrace the cloud to deliver their services, the security of these networks must remain a top priority. Approaches mentioned above are integral elements of a robust telco cloud security strategy. Telco operators must remain vigilant, adapt their security practices, and invest in cutting-edge technologies to protect their networks and services in the cloud. Safeguarding the future of telecommunications depends on robust, adaptive security measures.

6.5 Multi-cloud and Hybrid Cloud Approaches for the Telco Cloud

6.5.1 How multi-cloud and hybrid cloud align to SP Telco Cloud requirements

Multi-cloud involves the utilization of various cloud services offered by different cloud providers to store data, execute applications, and oversee workloads across diverse cloud platforms. In the telecommunications sector, "multi-cloud" signifies the utilization of multiple cloud service providers for the administration and operation of telecom infrastructure and services.

On the other hand, a hybrid cloud denotes a blended computing environment where applications are executed using a combination of computing resources, storage, and services across varying cloud environments. This encompasses both public (aka hyperscalers) and private clouds, including on-premises data centers and even edge locations.

Both multi-clouds and hybrid clouds offer great flexibility to Telcos. Multi-cloud helps to avoid vendor lock-in in cloud services and fosters efficient utilization of available cloud platforms as per the application requirements (latency, resources, etc.). Hybrid clouds help to keep the sensitive and secure workloads within the Telco's private cloud itself and at the same time provides scalability and resiliency for other workloads compatible with public cloud. These approaches help to increase the resilience and reliability of the Telco. There is lower risk of interruption as services are spread across different cloud service providers.

Telcos can establish a robust framework to strategically harness the security features of multiple cloud providers, effectively safeguarding against data breaches and cyber threats. The utilization of a multi-cloud approach empowers telecom firms to access a wider spectrum of cloud services and technologies. This, in turn, can expedite innovation and shorten the time required to introduce new products and services to the market.

Multi-clouds and hybrid clouds come with their own set of challenges. Attaining multi-cloud portability necessitates separating the procedures for deploying workloads and managing their lifecycle from being tied to a particular cloud technology. Telcos might need to develop cloud-agnostic tools to monitor workloads running on these different cloud stacks. A governance policy needs to be defined that details how the data could flow between different clouds without compromising on security aspects.

Not all applications are made same; thus, the best cloud solution will depend on factors like data throughput, latency, privacy demands, workload variations, geographic limitations, and cost-effectiveness. Managing these requirements call for an integrated multi-cloud strategy, which can streamline deployment and operation. It goes without saying that any Telco looking for robustness in their cloud-based network cannot work without a combination of multi-cloud and hybrid cloud approaches.

CHAPTER

7

Packet Core (5G NSA and SA) Design

7.1 Introduction

This chapter aims to give the reader a detailed view of the 5G core network and covers both non-standalone (NSA) and standalone (SA) versions.

Figure 7.1: 5G SA and NSA.

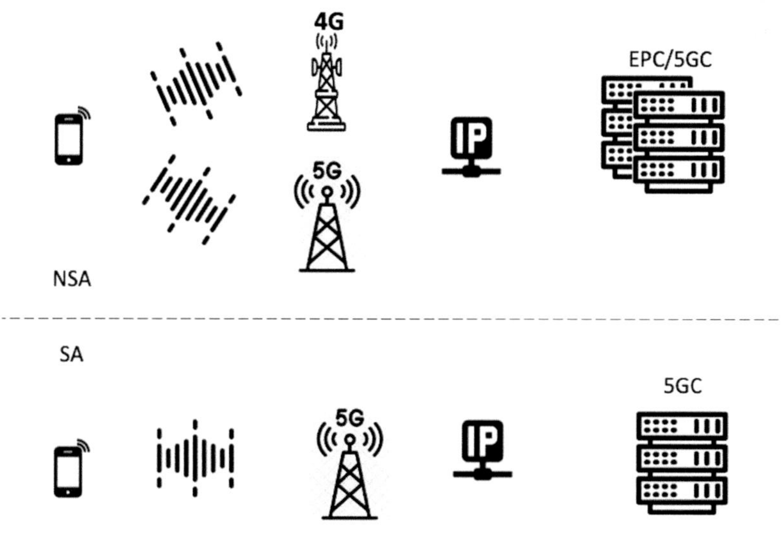

The 5G SA core (5GC) architecture aims to provide a flexible, scalable and efficient framework for supporting the diverse range of services and applications that 5G networks promise. It is designed to meet the evolving requirements of 5G use cases while enabling efficient network operations. The key targets of the 5G Core architecture are:

Enhanced mobile broadband (eMBB): 5GC aims to deliver significantly higher data rates, lower latency, and improved network capacity compared to previous generations. This supports applications such as ultra-high-definition video streaming, virtual reality, augmented reality, and other bandwidth-intensive services.

Ultra-reliable low-latency communications (URLLC): The architecture is designed to provide extremely low latency and high reliability, enabling applications that require real-time responsiveness and mission-critical reliability, such as industrial automation, autonomous vehicles, and remote surgery.

Massive machine-type communications (mMTC): The 5G core architecture is optimized to handle a massive number of connected devices, enabling the Internet of Things (IoT) and other applications that require efficient communication with a large number of devices, such as smart cities and smart agriculture.

Network slicing: Network slicing allows the 5G core to create virtual, independent, and customizable network instances within a single physical network infrastructure. This enables tailored network services for different use cases, optimizing resource allocation and ensuring service quality.

Service-based architecture (SBA): The 5G core introduces a service-based approach to network functions, enabling more flexible and dynamic service deployment, interaction, and composition. This makes it easier to develop and deploy new services rapidly.

Cloud-native and virtualization: The 5G core architecture leverages cloud-native principles and virtualization technologies to create more agile, scalable, and cost-efficient network solutions.

Automation and orchestration: Automation and orchestration capabilities are integral to the 5G core, allowing dynamic management of network resources, services, and functions. This helps optimize network operations and enhances the overall user experience.

Security and privacy: The architecture incorporates enhanced security mechanisms to address the increased threat landscape in 5G networks, ensuring the confidentiality, integrity, and availability of data and services. It also emphasizes user privacy and data protection.

Open interfaces and ecosystem: The 5G core architecture promotes the use of open, standardized interfaces to encourage innovation, interoperability, and the development of a vibrant ecosystem of applications and services.

Global roaming and interoperability: The 5G core architecture facilitates seamless international roaming and interoperability between different operators and networks, enabling a consistent user experience across different regions.

These targets collectively contribute to the overarching goal of the 5G core architecture, which is to provide a versatile and future-proof platform for enabling a wide range of use cases and services in the 5G era.

We look at the evolution of packet core from 4G to CUPS, followed by NSA and SA. Over the generations of packet core technologies, we have seen how each evolution helped in attaining objectives such as higher data rate, lower latency, varied use-cases, etc. The architectural elements of each technology are different, however, and are designed to work backwards to allow smooth transitions. The 3GPP standards have provided operators with multitude of options for evolving to or implementing a complete 5G network.

The NSA approach serves as an intermediate step towards achieving a 5G network. NSA allows the deployment of 5G New Radio (NR) to connect with a 4G core network. In this setup, the 5G radio relies on the 4G eNB for all control plane communication. The term "NSA" indicates that the 5G NR cannot function independently and requires the assistance of eNB for control plane signaling.

5G emphasizes the separation of control and data planes. This separation, known as control and user plane separation (CUPS), was introduced in the 4G core as well. Under CUPS, the legacy 4G SGW and PGW were divided into separate nodes for control and user planes. Consequently, both SGW and PGW, following CUPS, possessed their own control and user planes, which communicated through a well-defined interface.

When an operator transitions to the NSA option, the 5G gNB connects to the LTE eNodeB for all signaling, meaning that the LTE network controls the NR gNB. However, the data path remains separate, and the gNB establishes a direct S1-U tunnel with the SGW for all data traffic, encapsulated within the GTPU protocol. In NSA, only the access network operates in a 5G capacity, where the device communicates with the 5G NR, and the gNB directs the data traffic to the SGW.

As the operator moves to SA, it is expected to co-exist with NSA and legacy 4G networks. In an SA-based deployment, a 5G UE in the NR coverage area would connect to the gNB and subsequently to the core NFs such as SMF, UPF, AMF, etc. If a 5G UE moves to a 4G coverage area where 5G coverage is unavailable, it would still connect to 5G core (i.e., converged core with co-located SMF/PGW and AMF/MME) via LTE eNB. 3GPP has specified all aspects of this interworking. The most important feature of this interworking is that the UE's IP address remains the same, assuring session continuity as it shifts between the radio technologies.

7.2 Non-standalone (NSA)

To enable large-scale trials and deployments to begin early, 3GPP delivered an intermediate milestone in the form of "non-standalone" (NSA) 5G New Radio. This milestone involved utilizing the existing LTE radio and core network for functions like mobility management, session management, and macro-coverage, while introducing a 5G radio carrier to enhance subscriber throughput.

The 5G NSA primarily addresses the enhanced mobile broadband (eMBB) only while aspects like ultra reliable and low latency communication (URLLC) and massive machine type communications (MMTC) are addressed in the 5G standalone.

7.2.1 NSA architecture

Dual connectivity: Dual connectivity (DC) is employed when a device utilizes radio resources from different access points, such as an eNB or gNB. Furthermore, each network access point involved in the dual connectivity arrangement can have distinct roles (master/secondary). Using MR-DC, it is also possible to split the user plane bearers between the master and secondary RAN nodes. Any transfer of data between the nodes is based on a user plane connection across Xn/X2 interfaces. NR-DC (new radio dual connectivity) refers to dual connectivity between gNBs.

Multi-RAT dual connectivity: In multi-RAT dual connectivity (MR-DC) mode, the master RAN node refers to the eNB/gNB/ng-eNB responsible for terminating the core network connectivity. It serves as the anchor for mobility towards the core network. On the other hand, the secondary RAN node, which can be an eNB, ng-eNB, or gNB, operates in dual connectivity mode by providing supplementary radio resources to the device. However, it does not function as the master RAN node itself.

Under the generic umbrella of MR-DC, the following options are available for dual connectivity with multiple RATs (Figure 7.2).

EN-DC is used to denote E-UTRA-NR dual connectivity in option 3 deployments, where the eNB assumes the master role and the gNB serves as the secondary node.

NGEN-DC represents NG-RAN E-UTRA dual connectivity, where the ng-eNB serves as the master node and the gNB operates as the secondary node.

NE-DC stands for NR-E-UTRA dual connectivity, in which the gNB functions as the master node and the ng-eNB operates as the secondary node.

Figure 7.2: MR-DC options summary.

Name	Master	Secondary	Core
New Radio Dual Connectivity **(NR-DC)**	gNB	gNB	5GC
E-UTRA – NR Dual Connectivity **(EN-DC)**	eNB	en-gNB	EPC
NG-RAN – E-UTRA-NR Dual Connectivity **(NGEN-DC)**	ng-eNB	gNB	5GC
NR – E-UTRA Dual Connectivity **(NE-DC)**	gNB	en-gNB	5GC

Most operators have chosen the EN-DC option for NSA deployments with MR-DC. Figure 7.3 shows the EN-DC architecture.

Because the eNB functions as the master node, it enables S1-MME connectivity to the EPC (evolved packet core). Both the eNB and en-gNB (secondary gNB) have the capability to provide user plane connectivity through S1-U. The X2 reference point is utilized for controlling communication between the master and secondary node. Additionally, the X2-U reference point can be utilized between the eNB and en-gNB to facilitate data forwarding.

Figure 7.3: EN-DC option.

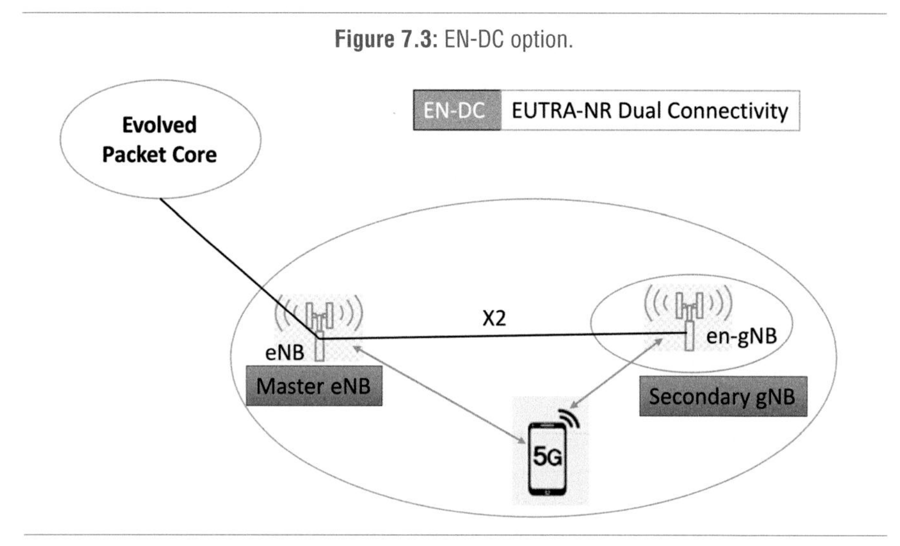

7.2.2 Procedures

7.2.2.1 EN-DC setup in RAN

Establishment of the X2 reference point between the master eNodeB and the secondary en-gNB is necessary before activating EN-DC. This establishment is accomplished through the EN-DC X2 setup procedure. It is important to note that either node can initiate this procedure. X2 application protocol (X2AP) messages are exchanged during this process. The EN-DC X2 setup procedure allows the two RAN nodes involved in EN-DC operation to exchange their cell configurations. Once the X2 setup is successfully completed, if a node needs to be added, modified, or deleted in a served cell, it will inform its peer node through the EN-DC configuration update procedure.

7.2.2.2 Security and initial attach

EN-DC configuration can only take place once security has been activated at the master eNB. In terms of key distribution, bearers that terminate at the master eNB will require the usual KeNB. However, for bearers that terminate at the secondary en-gNB, a specific security key called S-KgNB will be provided.

The NAS attach request message will contain the UE additional security capability information element which lists the 5GS encryption and integrity checking algorithms that the device supports. If the device supports 5G-EA0 to 5G-EA3, it must also support the equivalent algorithms in the E-UTRAN. The same rule applies to the integrity algorithms.

Establishment of EN-DC connection requires the function to be supported by both UE and the network. The UE signals it's support for EN-DC by setting the "dual connectivity with NR" flag in the "UE network capability IE" sent in the NAS attach request message. On the other side, MME can decide to accept or reject EN-DC for this UE by appropriately setting the "use of dual connectivity with NR is restricted" flag in the attach accept message.

Activation of EN-DC occurs after the device has attached to the network, as the security at the master eNB needs to be established. MME informs the eNB that the device supports EN-DC during the attach process via the "UE radio access capability information" IE in the S1AP Initial UE context setup message, which contains a flag for EN-DC support. The eNB will attempt to establish EN-DC operation immediately if possible. (Availability of feasible en-gNB, cell restrictions, etc.)

7.2.2.3 Secondary node addition

After the UE successfully performs the attach procedure, EN-DC will be activated resulting in addition of the secondary node (en-gNB).

The X2AP secondary gNB addition request message is sent by eNB to en-gNB which contains the information to setup the X2AP connection for the specific UE. The master eNB provides the NR security capabilities and keys to the en-gNB. It also sends the bearer information such as E-RABs to be added, UE AMBR, TEIDs, etc. In response, the en-gNB sends the admitted E-RAB list back.

7.2.2.4 RRC reconfiguration and user plane path switch

Once EN-DC is setup in the RAN, the UE needs to be updated via the RRC reconfiguration procedure to use the secondary node. The master eNB sends an RRC NR configuration message to the UE following which it will be able to access the secondary en-gNB. The response from the UE (RRC NR response) will be transferred by eNB to en-gNB to complete the process. The UE will perform the random access (RACH) procedure to access the en-gNB.

It is possible that during the secondary node addition process, a data forwarding path is established to allow traffic to flow. To allow subscriber traffic to go to the core via the secondary node, the user plane path should be updated.

The master eNB initiates this procedure by sending a S1AP E-RAB modification message containing the GTP TEID of the secondary en-gNB. In the core, MME will send the TEID to SGW via the GTPv2 modify bearer request.

7.2.3 Adaptations in core network for NSA

As NSA supports higher throughput compared to 4G, some adaptations are needed in the 4G core to support enhanced bit rates.

7.2.3.1 QOS

5G NR provides a maximum downlink data speed of 20 Gbps and an uplink data speed of 10 Gbps. Certain interfaces within the core are equipped to handle the encoding and decoding of this 5G data throughput. For instance, the NAS interface can support speeds of up to 65.2 Gbps (APN-AMBR), while the S5/S8/S10/S3 interfaces (GTP-v2) can handle up to 4.2 Tbps. However, the diameter interfaces S6a and Gx are limited to a maximum throughput of 4.2 Gbps, the S1-AP interface supports only up to 10 Gbps, and NAS supports up to 10 Gbps (MBR, GBR).

To accommodate the 5G data speeds, new attribute value pair (AVP)/information elements (IE) have been introduced in the S6a, Gx, S1-AP, and NAS interfaces. Extended-max-requested-BW-UL and extended-max-requested-BW-DL AVPs are implemented in HSS – in the customer profile. MME will use these values from HSS in the existing APN-AMBR IE to setup the bearer towards SGW/PGW. On the PGW, for policy authorization, new AVPs extended-APN-AMBR-UL and extended-APN-AMBR-DL are used to authorize the values. PCRF could approve or modify the received values. PCRF provided values are communicated back to MME by PGW and ultimately MME sends them to eNB/UE via an extended UE aggregate maximum bit rate downlink and extended UE aggregate maximum bit rate uplink (towards eNB) and extended APN aggregate maximum bit rate (towards UE)

URLCC QCI 80 (non-GBR resource type), QCI 82 and 83 (GBR resource type) are also added into the supported QCIs list by SGW/PGW.

7.2.3.2 Charging

The SGW/PGW is enabled to collect and report the secondary RAT type and traffic volume over that RAT for offline charging. The data collection is achieved by the eNodeB reporting the usage information when an eNodeB stops serving a UE in connected state. MME forwards a received volume report via S11 signaling to the SGW that handled the traffic. This reporting can be periodic as well.

SGW can receive a secondary RAT usage data report from MME over the S11 interface in messages such as create-session-request, modify-bearer-request, delete-session-request, etc. For periodic usage reporting, SGW can receive periodic secondary RAT usage data report from MME over S11 interface in the change-notification-request message.

PGW can receive a secondary RAT usage data report from SGW over the S5/S8 interface in the following messages: modify-bearer-request, delete-session-request, delete-bearer-response, delete-bearer-command.

For periodic usage reporting, PGW can receive a periodic secondary RAT usage data report from SGW over the S5/S8 interface in a change-notification-request message. SGW and PGW can add the secondary RAT usage data in the CDRs.

7.2.4 Control user plane separation (CUPS)

In the core network, the traffic includes signaling and data. Signaling traffic is very minimal as compared to the data traffic. CUPS intends to separate the signaling processing and data traffic processing. The gains are high with this separation. Depending on operator's traffic profile, the control and user plane nodes can scale independently. Capacity enhancement can be done without touching the control plane – manually or even automatically via orchestration.

In 4G, CUPS involves splitting each of the SGW and PGW nodes into individual control (SGW-C, PGW-C) and user plane (SGW-U, PGW-U) entities. They will exchange messages via a new protocol called PCFP. This split means that some of the tasks existing in a PGW is now separated. Traffic processing, deep packet inspection, usage monitoring, policy enforcement, etc. are done on the user plane node whereas control plane node is restricted to pure signaling.

CUPS gives a lot of flexibility to the operator. Throughput capacity can be scaled up by adding only user plane nodes, saving unnecessary resources wastage on the control plane.

7.2.4.1 CUPS advantages

- CUPS is the first step towards 5G. It aligns with the basic 5G premise of separating the control and user planes.

- CUPS allows independent and on-demand scaling of control and user plane nodes. This adds scope for load-based orchestration to increase or decrease capacity without impacting the service.
- Dedicated user planes allow them to be placed closer to the radio – helping in reducing latency and offloading heavy (e.g., video) traffic from the backbone. MEC use-cases can be implemented via dedicated user plane nodes.
- Centralized control plane helps management of functions like billing, policy enforcement, signaling monitoring, etc. easier.
- Limited impact to the system on single user plane node failure.

7.2.4.2 CUPS architecture

CUPS separates the control plane and user plane functionalities of the packet core network, which is responsible for routing and delivering data packets in the mobile network. CUPS introduces a separation between these planes, allowing them to be deployed and scaled independently. This separation provides flexibility, scalability, and improved resource utilization in the mobile network.

Figure 7.4: CUPS node layout.

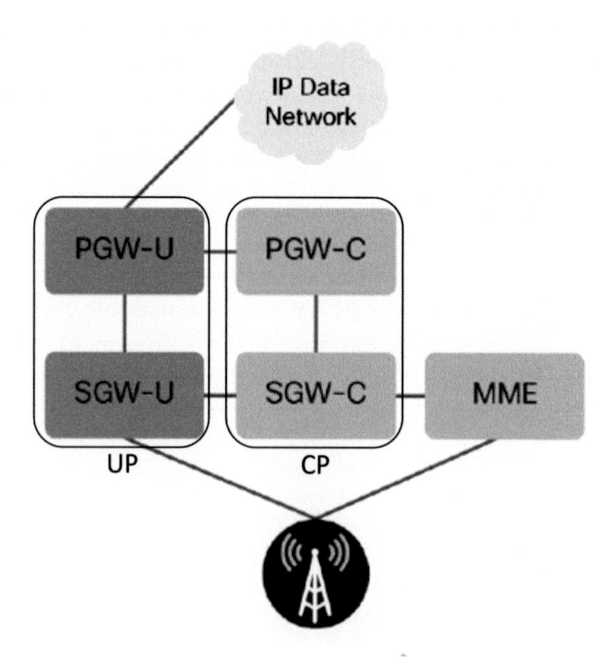

In the traditional architecture, the control plane and user plane functionalities are closely integrated within the same network nodes. The CUPS architecture addresses this by centralizing the control plane functions in dedicated control plane nodes (CP) while distributing the user plane functions in user plane nodes (UP). The separation allows for better scalability, as additional user plane nodes can be added to handle increasing data traffic without affecting the control plane.

The communication between the control plane and user plane nodes is done using standard protocol – packet forwarding control protocol (PFCP). The control plane nodes send instructions and policies to the user plane nodes, which then handle the data traffic accordingly (Figure 7.4).

7.3 Standalone (SA)

The 5G standalone core (SA) is the heart of the next generation 5G network, providing the essential framework that supports the advanced capabilities and features of 5G technology. It represents a significant departure from previous generations of mobile network architecture by introducing a more flexible, scalable, and efficient design that caters to a diverse range of services and applications.

Unlike NSA, which relied on the existing 4G infrastructure, the 5G SA core is purpose-built to fully harness the capabilities of 5G. It forms the backbone of the standalone 5G network, enabling the network to operate independently of legacy technologies. This independence empowers the SA core to deliver an array of new use cases without being overburdened by the compatibility to previous generations.

7.3.1 Service based architecture (SBA)

Service-based architecture (SBA) is a fundamental architectural concept in the 5G core that brings benefits to the deployment, operation, and evolution of 5G networks.

Legacy core is based on direct connections between two network components using point-to-point interfaces. These interfaces are mostly very different, encompassing differences not just in the configuration of the application layer but even in the underlying transport framework. This conventional approach has resulted in complex architectures that demand significant effort to expand or alter functionalities, hindering the seamless sharing of functions throughout the system.

The basic premise in SBA is that each NF offers a set of services. An NF service represents a particular capability of a specific network function. This capability is made accessible by the network function (acting as the NF service producer) through a service-based interface, which is then utilized by authorized network functions (serving as NF service consumers). An NF has the capacity to present multiple NF services, just as it can utilize several NF services. Network functions possess varying capabilities, thereby offering diverse NF services tailored for different consumers. Every NF service must be reachable through an interface, with this interface potentially comprising one or multiple operations.

In the 4G core, each interface has specific protocol associated – such as GTPv2, diameter, radius, etc. However, in 5G core the corresponding interfaces now use SBA. There is no protocol specific to an interface.

The protocol framework for SBA consists of:

- HTTP/2 as application layer
- TCP as transport layer protocol
- The JavaScript Object Notation (JSON) for data serialization
- RESTful (representational state transfer) framework for the API design
- OpenAPI 3.0.0 as the interface definition language.

SBA uses two mechanisms for message transfer between the NFs:

Request-response: An NF_B (acting as the provider of an NF service) is prompted by another control plane NF_A (serving as the recipient of the NF service) to deliver a specific NF service. This service can involve executing an operation, supplying information, or both. NF_B carries out the requested NF service as per NF_A's solicitation. To fulfil this request, NF_B might also engage other NFs to utilize their respective NF services. In the request-response mechanism, interaction occurs on a one-to-one basis between the service consumer and the service producer.

Subscribe-notify: Within the control plane, an NF_A (functioning as the user of the NF service) subscribes to an NF service provided by another control plane NF_B (acting as the provider of the NF service). Multiple control plane NFs could potentially subscribe to the same NF service from NF_B. Subsequently, NF_B informs all the subscribed NFs about the outcomes of the respective NF service.

The NF service authorization framework ensures that the NF service consumer is authorized to access the NF service provided by the NF service provider, according to the policies of the NF and the policies of the serving operator.

NF registration and de-registration is supported via the NRF (network repository function). NRF holds records of accessible instances of network functions (NFs) along with the services they can offer. Each NF instance enrolls with the NRF, listing all the NF services it can provide. As part of this registration process, the NF instance shares its NF profile, which the NRF maintains. When the NF instance is preparing to shut down or deliberately disconnect from the network in an orderly fashion, it de-registers from the NRF.

To summarize the advantages, SBA offers the following:

Flexibility and agility: SBA introduces a modular and service-oriented approach to network design. In the 5G core, network functions are decoupled into independent services that communicate with each other using standardized interfaces. This modular structure allows for easier development, deployment, and modification of network services. New services can be added or updated without affecting the entire network.

Efficient resource utilization: With SBA, resources can be allocated more efficiently based on actual demand. Network functions can be dynamically instantiated, scaled, and terminated as needed, optimizing resource usage and reducing operational costs.

Service orchestration and composition: SBA enables the orchestration and composition of services, allowing operators to create tailored services by combining various network functions. This feature supports the creation of network slices – virtual networks customized for specific use cases with distinct performance characteristics.

Interoperability and interconnection: Standardized service interfaces in SBA foster interoperability and seamless interaction between different network components, even if they come from different vendors. This promotes healthy competition, accelerates innovation, and prevents vendor lock-in. It also facilitates multi-operator scenarios and the integration of third-party services into the network.

Futureproofing and evolution: SBA makes it easier to introduce new technologies and functionalities to the network over time. As new services and applications emerge, they can be integrated into the network through well-defined service interfaces.

Easier maintenance and troubleshooting: With a modular architecture and clear service boundaries, network maintenance and troubleshooting become more straightforward. Issues can be isolated to specific services, minimizing the impact on the overall network, and reducing downtime.

7.4 5G Core Network Functions

Following are the main 5G NFs as defined by 3GPP and shown in Figure 7.5:

Figure 7.5: 5G Network functions and interface reference points.

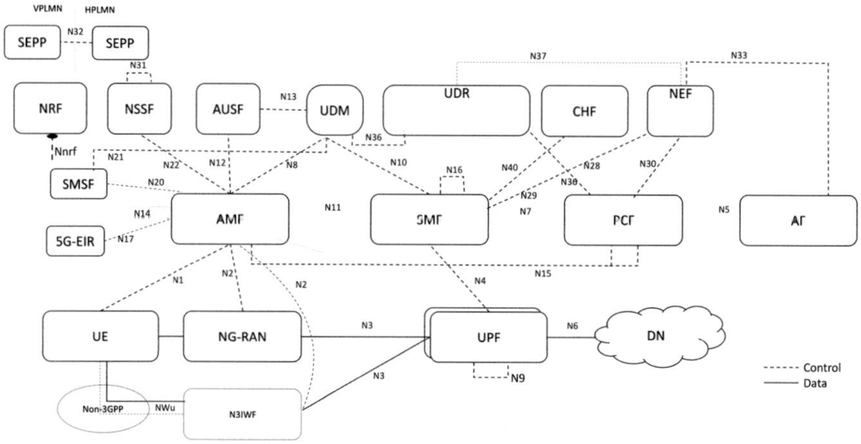

- Session management function (SMF)
- Access and mobility management function (AMF)
- User plane function (UPF)
- Policy control function (PCF)
- Charging function (CHF)
- Authentication server function (AUSF)
- Unified data management (UDM)
- Unified data repository (UDR)
- Network repository function (NRF)
- Network slice selection function (NSSF)

- Network exposure function (NEF)
- Security edge protection proxy (SEPP)
- 5G-equipment identity register (5G-EIR)
- SMS function (SMSF)
- Application function (AF)
- N3 interworking function (N3IWF).

7.4.1 SMF

The session management function (SMF) in a 5G network is responsible for various technical functions that enable the establishment, control, and management of user sessions. The main functions of SMF are:

Session establishment and control: The SMF is responsible for setting up and controlling user sessions. It also manages modifications during active sessions, and orchestrates the release of sessions.

IP address allocation: SMF allocates IP addresses to user equipment (UE) for IP packet data unit (PDU) sessions. This ensures that the UE has unique identifiers and can communicate effectively with data networks.

Session continuity: SMF manages service and session continuity. It provides mechanisms to ensure that ongoing sessions are not disrupted when UE moves between different network cells or access points.

User plane management: SMF oversees the user plane for data connectivity, ensuring efficient data transmission between the UE and the data networks. It controls traffic steering and quality of service (QoS) for individual sessions.

Interactions with AMF: SMF communicates indirectly with UEs through the access and mobility management function (AMF). It relays session-related messages between the UE and the SMF to facilitate session setup and management.

Policy enforcement: The SMF interacts with the policy control function (PCF) to retrieve policies. These policies guide the configuration of the user plane function (UPF) for PDU sessions, including traffic routing and QoS enforcement.

Charging and billing: SMF collects charging data and controls charging functionality in the UPF. It supports both offline and online charging models, ensuring accurate billing for the services used by UEs.

Overall, the SMF plays a central role in ensuring the efficient and reliable operation of 5G networks by managing user sessions, IP address allocation, session continuity, and quality of service, while also facilitating charging and billing processes.

7.4.2 AMF

The access and mobility management function (AMF) is responsible for functions that handle mainly the mobility and security in 5G core.

Mobility management: The AMF plays a pivotal role in mobility management within 5G networks. It enables UEs to move between different network cells and access points while maintaining continuous connectivity. This includes managing handovers and ensuring uninterrupted service.

Security: The AMF is responsible for user authentication, authorization, and the enforcement of security protocols. It ensures the confidentiality and integrity of data transmission by establishing secure communication channels.

Policy enforcement: The AMF interacts with the policy control function (PCF) to enforce policies related to network access, quality of service (QoS), and user-specific preferences. This ensures that each piece of UE receives the appropriate level of service based on its profile and subscription.

Interactions with the SMF: The AMF communicates indirectly with the session management function (SMF), relaying session-related messages between UE and the SMF. This coordination is crucial for session setup and management.

Subscriber location tracking: AMF keeps track of the location of UEs within the network, allowing for efficient routing and management of traffic.

Service continuity: In cases of service disruption, the AMF facilitates service continuity by managing procedures like idle mode mobility and registration.

Authentication and key management: The AMF handles authentication and key management functions, ensuring secure communication between UE and the network.

The AMF is a fundamental element of the 5G network, responsible for mobility management, security, policy enforcement, and ensuring a seamless and secure user experience.

7.4.3 UPF

The user plane function (UPF) is responsible for efficient routing and forwarding of user data traffic. User data traverses only UPF in the 5G core.

Packet routing and forwarding: The UPF serves as a packet router and forwarder. It receives user data packets from gNodeBs and directs them to their intended destinations (internet, MEC, etc.) based on the appropriate routing rules and policies.

Quality of service (QoS) enforcement: The UPF enforces QoS policies to ensure that different types of traffic receive the appropriate level of service quality. It manages bandwidth allocation, latency, and packet prioritization based on the QoS requirements specified for each user or application.

Traffic steering: The UPF can perform traffic steering to optimize the path of user data traffic. This may involve selecting the most suitable data path based on factors such as network conditions, load balancing, and service requirements.

Data traffic optimization: The UPF can perform data optimization tasks, including data compression, deduplication, and header compression, to reduce bandwidth consumption and improve network efficiency.

User data charging: The UPF collects information related to data usage and contribute to the charging process by providing data consumption details to the charging function (CHF) via SMF for billing purposes.

Network function chaining (NFC): The UPF may be involved in network function chaining, which allows the creation of service chains by orchestrating different network functions in a specific order to process user data traffic.

7.4.4 PCF

The policy control function (PCF) serves several important functions within the context of session management, access and mobility control, UE access selection, PDU session selection, etc.

The PCF provides the SMF with policies related to quality of service (QoS), charging, and traffic forwarding for the data flowing in the subscriber PDU session. The PCF can also guide the SMF to choose a user plane according to specific policies set for traffic type or subscriber.

The PCF policies sent to the AMF may contain service area restrictions for the UE or access type priority or even RFSP (radio frequency selection priority). The gNodeB can match the RFSP index to its local configuration to apply specific radio resource management (RRM) policies, such as cell reselection or frequency layer redirection.

UE policies related to discovery and selection of non-3GPP networks (ANDSP) or session continuity mode selection, network slice selection, data network name selection (collectively URSP), etc. can be send from the PCF via the AMF.

7.4.5 CHF

The charging function (CHF) is a unified entity performing both online and offline charging in the 5G core. CHF interfaces with SMF to collect usage monitoring data and creates CDRs for retrieval by billing system. The CHF interfaces with PCF to update about the usage, thus empowering PCF to enforce appropriate policies based on the usage and allowed spending limits of the subscriber.

7.4.6 AUSF, UDM, UDR

The authentication server function (AUSF) performs authentication and key agreement procedures to establish mutual authentication between the UE and the network. It also provides security parameters to protect information in the UE update procedure. The AUSF is always located in the home PLMN of the subscriber.

The unified data management (UDM) is always located in the home PLMN of the UE. The UDM uses subscription data stored in the UDR to execute processes like authentication/authorization of access/services, registration management and reachability for terminating services. The UDM generates the authentication credentials that the AUSF uses to authenticate UE. The UDM also keeps track of the serving AMF for UE and the serving SMF/SMFs for the PDU sessions of the UE.

The unified data repository (UDR) is the database that stores the subscription data and other information related to network or user policies.

The UDR allows retrieval of subscription data (by the UDM), policy data (by the PCF) and structured data for exposure (by the NEF). During roaming, the UDR in home PLMN and the UDR in visited PLMN may be used to serve UE, with subscription data from the home UDR and policy data from the visited UDR.

7.4.7 NRF

The network repository function (NRF) plays an important role in SBA. It is the repository of the profiles of the network functions that are available in the network. It allows service consumer to discover and select suitable service producers dynamically. The NF profiles can be modified any time to reflect the latest changes in the NFs. The NRF also checks for liveliness of the NF to ensure that only working NF producer details are sent to the NF consumer. The NF profile contains information like NF type, address, capacity, supported NF services and addresses for each NF service instance. The NRF provides profiles only to a consumer who has the authority to discover that specific NF service.

7.4.8 NSSF

The network slice selection function (NSSF) has information on all available slices in the network. It selects the network slice(s) to be used for UE, based on the requested and subscribed S-NSSAI values. NSSF also selects the set of AMFs that should serve the UE. The AMF may be dedicated to one or a set of network slices. Note that each PLMN has its own NSSF.

7.4.9 NEF

The network exposure function (NEF) supports monitoring and exposure of event and capabilities from the 5G core towards applications and network functions inside and outside the operator's network.

7.4.10 SEPP

The security edge protection proxy (SEPP) is a non-transparent proxy that resides at the boundary of different PLMN's 5G core networks and provides message filtering and policing on the control plane interfaces (service producers and consumers in different PLMNs). It also helps in topology hiding.

7.4.11 5G-EIR

The 5G-EIR helps to check if the permanent equipment ID (similar to IMEI)) has been blacklisted or not. This can be used by operators to block access to the network if the device has been stolen and blacklisted.

7.4.12 SMSF

The SMS function (SMSF) connects the legacy SMS nodes to the AMF. The SMSF is responsible for SMS authorization, SMS relay and protocols, SMS charging, etc. SMS authorization checks the UE requests regarding SMS related subscription data retrieved from the UDM. The SMS relay consists of transferring SMS, via AMF, from UE to SMS-GW and vice versa.

7.4.13 AF

The application function (AF) is a network function which interacts with the 3GPP core network, however is not specified by 3GPP. Depending on the operator, an AF might be considered as a trusted entity and can interact directly with the 3GPP NFs without going via the NEF.

7.4.14 N3IWF

The N3 interworking function (N3IWF) is the network element that serves a gateway or interworking function that enables communication between 5G networks and non-3GPP networks. Non-3GPP networks can include technologies like Wi-Fi, Ethernet, etc. N3IWF supports IPsec tunnel establishment with the UE using IKEv2/IPsec protocols. Separate tunnels for NAS signaling messages and user data traffic are created with UE. N3IWF has the NGAP interface (over SCTP) towards AMF for signaling and GTP-U endpoint to carry the data traffic towards UPF.

7.4.15 Interface references and protocols in 5G core

Figure 7.5 also shows the interface reference names in 5G.

The protocol stack between the interfaces are shown in the Figures 7.6–10.

Figure 7.6: Control plane protocol stack 1.

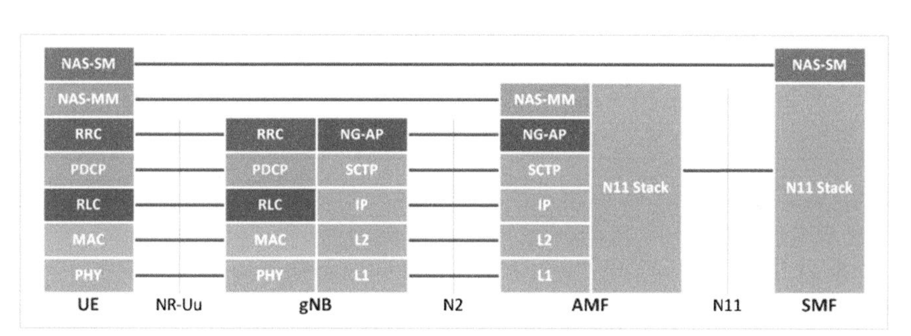

Figure 7.7: Control plane protocol stack 2 (inter-gNB and SMF/UPF).

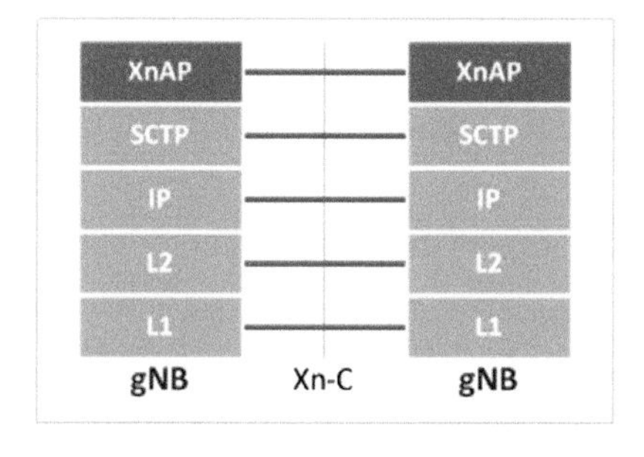

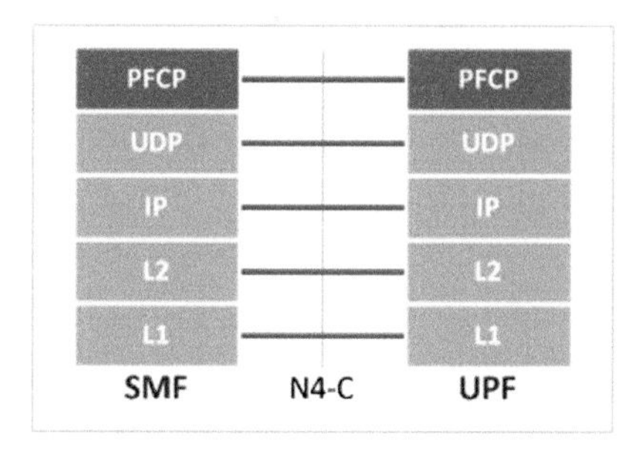

The SBI stack is used in all SBA interfaces. Note that TLS is optional and can be used to enhance security in the SBI stack (secure HTTP).

Between the gNodeB and AMF (N2), the NGAP (NG application protocol) is defined. Similarly, PFCP (packet forwarding control protocol) is defined on the N4 interface between SMF and UPF. GTP-U is carried forward from 4G for encapsulation and transport of the data traffic in the user-plane.

Figure 7.8: Service based interfaces stack.

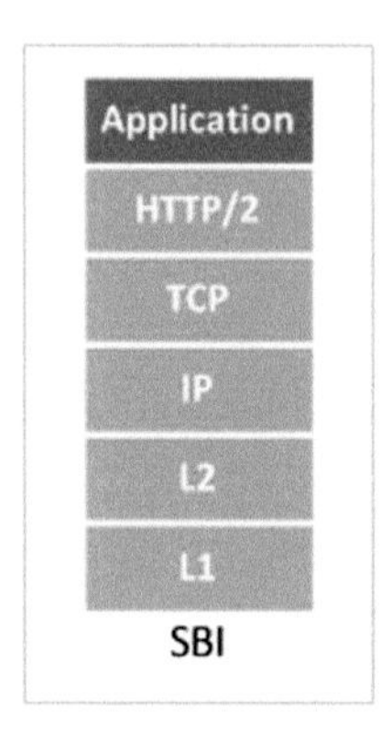

Figure 7.9: User plane protocol stack 1.

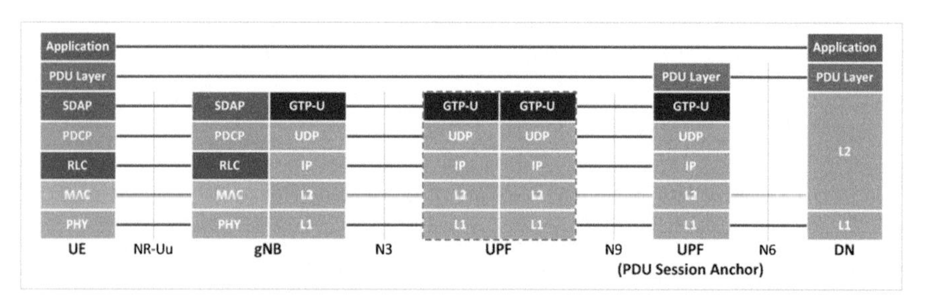

Figure 7.10: User plane protocol stack 2 (inter-gNB and SMF/UPF).

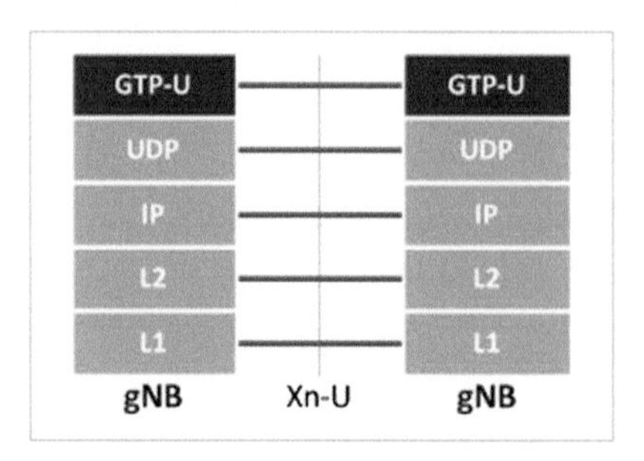

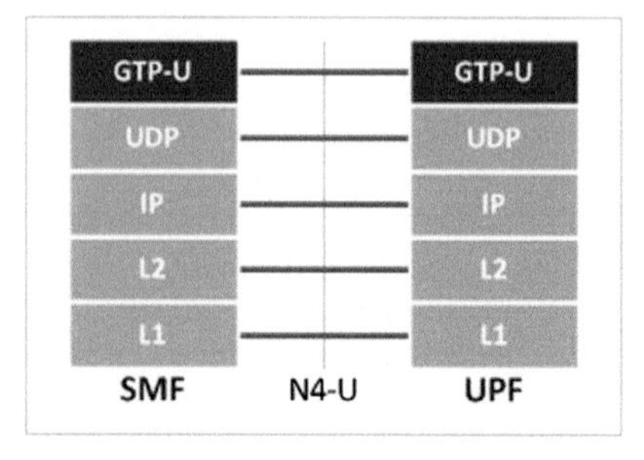

7.5 5G Core Concepts

In this section we explore the different procedures in the 5G core happening between the 5G NFs described in the previous chapter.

7.5.1 Identities in 5G

5G identities can be classified as subscriber related and network related.

Subscription permanent identifier (SUPI): SUPI is the globally unique identifier in 5G systems, typically in the form of the traditional IMSI or NAI (network access identifier). SUPI is never sent clearly across radio and is encrypted using a public key to form SUCI (subscription concealed identifier). Within the core, the SIDF (subscriber identifier de-concealing function), decrypts the SUCI to retrieve the SUPI using the private key. Using IMSI as the 5G SUPI helps in easier backward compatibility with 4G EPC core.

5G-globally unique temporary identifier (5G-GUTI): The AMF allocates this temporary ID during registration to keep the SUPI confidential and uses 5G-GUTI for all exchanges over the radio. GUTI comprises of the GUAMI (globally unique AMF ID) and the 5G-TMSI (5G temporary mobile subscriber identity). A shorter version of GUTI known as 5G-S-TMSI is used to optimize paging.

Permanent equipment identifier: The PEI is associated with the physical UE. The PEI in the form of the IMEI (international mobile equipment identity) is used if the device supports 3GPP access technology.

7.5.2 QOS in 5G

Quality of service is the ability of the system to provide different treatments for the data packets based on type of packet, subscriber, application, etc. The treatment here can refer to providing assured throughput, prioritization during congestion or even guaranteed level of performance of the data flow.

In 5G, the basic unit is the QOS flow. A given PDU session can have multiple QOS flows, each with its own different QOS properties. The QOS flow identifier (QFI) identifies each flow with its own QOS properties within the PDU session (Figure 7.11). The SMF determines which QFI a particular SDF (service data flow) must use. On the device side, classification is based on QOS rules received during the PDU session.

Establishment or modification: At the UPF, classification is based on a PDR (packet detection rule), which has all the values mapped from the QOS rule.

Figure 7.11: QFIs in 5G.

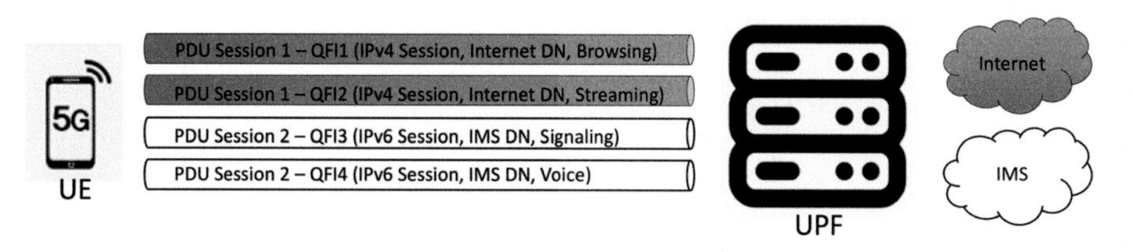

7.5.2.1 QOS properties

The QOS parameters in 5G are evolved from those of 4G. The following are the QOS parameters:

- 5QI
- Allocation and retention priority (ARP)
- UE/session AMBR
- Reflective QOS (newly introduced in 5G)
- Guaranteed flow bit rate
- Maximum flow bit rate.

Like QCI in 4G, 5QI is used to determine QOS characteristics attributed to a QOS flow. These characteristics are:

Resource type: GBR, non-GBR or delay critical.

Priority level: Traffic handling priority for the flow.

Packet delay budget: Upper limit (ms) for packet delay between UE and UPF.

Packet error rate: Upper limit for packet loss.

Averaging window: Period over which the GFBR/MFBR are calculated for GBR flows.

Maximum data burst volume: Largest amount of data that the 5G access network is expected to handle within a period of the packet delay budget.

3GPP has defined standard values for the above characteristics and has grouped them based on traffic type and requirements. If the service provider uses standard values, just the 5QI needs to be communicated and each node will apply the QOS characteristics that correspond to that 5QI.

Table 7.1 shows the standard values corresponding to the 5QI values as per 3GPP TS 23.501.

Table 7.1: Standardized 5QI to QoS characteristics mapping.

5QI value	Resource type	Default priority level	Packet delay budget	Packet error Rate	Default maximum data burst volume	Default Averaging window	Example services
1	GBR	20	100 ms	10^{-2}	N/A	2000 ms	Conversational voice
2		40	150 ms	10^{-3}	N/A	2000 ms	Conversational video (live streaming)
3		30	50 ms	10^{-3}	N/A	2000 ms	Real time gaming, V2X messages Electricity distribution – medium voltage
4		50	300 ms	10^{-6}	N/A	2000 ms	Non-conversational video (buffered streaming)
65		7	75 ms	10^{-2}	N/A	2000 ms	Mission critical user plane push to talk voice
66		20	100 ms	10^{-2}	N/A	2000 ms	Non-mission-critical user plane push to talk voice
67		15	100 ms	10^{-3}	N/A	2000 ms	Mission critical video user plane
71		56	150 ms	10^{-6}	N/A	2000 ms	“Live” uplink streaming
72		56	300 ms	10^{-4}	N/A	2000 ms	“Live” uplink streaming

Table 7.1: Continued.

5QI value	Resource type	Default priority level	Packet delay budget	Packet error Rate	Default maximum data burst volume	Default Averaging window	Example services
73		56	300 ms	10^{-8}	N/A	2000 ms	"Live" uplink streaming
74		56	500 ms	10^{-8}	N/A	2000 ms	"Live" uplink streaming
76		56	500 ms	10^{-4}	N/A	2000 ms	"Live" uplink streaming
5		10	100 ms	10^{-6}	N/A	N/A	IMS signaling
6		60	300 ms	10^{-6}	N/A	N/A	Video (buffered streaming)TCP-based (e.g., www, e-mail, chat, ftp, etc.)
7	Non-GBR	70	100 ms	10^{-3}	N/A	N/A	Voice, video (live streaming)interactive gaming
8		80	300 ms	10^{-6}	N/A	N/A	Video (buffered streaming)TCP-based (e.g., www, e-mail, chat, ftp, etc)
9		90					
69		5	60 ms	10^{-6}	N/A	N/A	Mission critical delay sensitive signaling
70		55	200 ms	10^{-6}	N/A	N/A	Mission Critical Data
79		65	50 ms	10^{-2}	N/A	N/A	V2X messages
80		68	10 ms	10^{-6}	N/A	N/A	Low latency eMBB applications augmented reality
82	Delay critical GBR	19	10 ms	10^{-4}	255 bytes	2000 ms	Discrete automation

Table 7.1: Continued.

5QI value	Resource type	Default priority level	Packet delay budget	Packet error Rate	Default maximum data burst volume	Default Averaging window	Example services
83		22	10 ms	10^{-4}	1354 bytes	2000 ms	Discrete automation V2X messages
84		24	30 ms	10^{-5}	1354 bytes	2000 ms	Intelligent transport systems
85		21	5 ms	10^{-5}	255 bytes	2000 ms	Electricity distribution- high voltage, V2X messages
86		18	5 ms	10^{-4}	1354 bytes	2000 ms	V2X messages (advanced driving: collision avoidance, platooning with high LoA

7.5.2.2 Reflective QOS

Reflective QoS is newly introduced in 5G. When the RQA (reflective QoS attribute) is utilized, the UE will derive the QOS rule from the downlink packets it is receiving (the packet filter and the QFI). Thereafter, all uplink traffic that meets the same packet filter will receive the same QOS treatment by the device. This is applicable only to non-GBR QOS. SMF, UPF, and gNodeB manages the allocation and application of reflective QOS on the device.

7.5.3 NF selection

During the core procedures, NFs need to be selected by other NFs. Depending on the type of NF, different selection strategy is used. For example, an AMF selection by gNodeB will be based on the availability of the previous AMF, load balancing and network slice. The SMF selection by the AMF for PDU session establishment can be based on local configuration or via NRF taking into account the data network, SMF load, network slice, etc. The UPF selection by the SMF can again be based on load, data network, slice, location of subscriber or based on policies received from the PCF.

7.5.4 Network registration

UE has to perform the registration before setting up the PDU session. In 4G the PDN session establishment is part of registration. However, in 5G they are independent. The following are the main steps in initial network registration:

- RRC connection establishment
- Registration request from UE
- AMF selection by gNodeB
- Identity retrieval by AMF
- Subscriber/network authentication via AUSF
- Subscription data retrieval from UDM
- Allowed slice information retrieval from NSSF and AMF redirection if required
- Access management policy retrieval from PCF
- Activation of any pre-existing PDU sessions
- Registration accept to UE.

7.5.5 PDU session establishment

The PDU session provides an association between the UE and a specific data network. The PDU session is established on demand based on UE configuration and URSP policies. Data traffic flows only after the PDU session is established. The main steps are:

- PDU session establishment request from UE
- SMF selection by AMF
- Subscription data retrieval from UDM by SMF
- Session management policy retrieval from PCF
- UPF selection and N4 session establishment between SMF/UPF
- PDU session establishment accept to UE providing N3 tunnel information on UPF side
- N3 tunnel allocation on gNodeB side and update of the same towards AMF/SMF/UPF
- Transfer of UL/DL data.

7.5.6 Mobility

Depending on the registration state and connection management state, there are different mobility procedures in the 5G core. Registration states are RM-registered and RM-deregistered, whereas the CM states are CM-connected and

CM-idle. The RRC states of the device in 5G are RRC-idle (in CM-idle) and RRC-inactive/RRC-connected (in CM-connected).

Cell reselection happens in CM-idle mode so that the device is always camped on the cell with the best signal coverage. The device performs periodic measurements to determine if reselection is required.

To keep the network updated of the current location of the device in CM-idle mode, it performs a periodic registration area update. This can be periodic or during a change of registration area (tracking area). In addition, depending on the device type (IOT or smartphone), discontinuous reception mode can be activated, wherein the device (mostly IOT devices) can go to deep sleep for longer time to save battery. It will wake up for the periodic updates and will not be reachable otherwise.

In CM-connected mode, handovers happen where the network chooses the target cell. Depending on the type of interface available, the handovers are classified as Xn and N2. Xn HO is used when the Xn interface is available between the gNodeBs. The handover is done in different phases – measurement, HO preparation, HO execution and path switch. In Xn HO, AMF is involved only in the path switch step. When the Xn interface is not present, AMF co-ordinates the handover process over the N2 interface and this is the N2 HO.

7.5.7 Interworking with 4G

Depending on the capability of the device and availability of the N26 interface (between AMF and MME), the device can be in single registration mode (either 4G or 5G) or dual registration mode (both 4G and 5G). If the N26 interface is available, the device will be in single registration mode. In this case, the AMF/MME can retrieve the mobility and session management information from the peer node to proceed with registration, security and PDU/PDN session establishment procedures. Note that the session anchor point remains same – combined SMF/PGW.

When N26 is not available and the device is in single registration mode, it has to perform the attach and PDN establishment (5G to 4G) or registration and PDU establishment (4G to 5G) based on the direction of handover. It is possible that the IP address is not preserved during this handover. If the device is in dual registration mode, it can pre-register in the target network and move the required PDU/PDN sessions only. Note that in this case, the device can choose which bearers need to be moved to target network and need to be retained in the current network.

7.5.8 Interworking with non-3GPP networks

N3IWF forms the integral part for interconnecting the non-3gpp access with 5G core. It performs a similar function to the ePDG in 4G for untrusted access. Devices can simultaneously attach to both 5G and non-3GPP networks at the same time, implying that there will be a separate registration management context in the AMF for the two networks. The device performs the registration to 5G core starting with establishment of the IPSec SA with N3IWF. During this procedure, IKE signaling will be used for EAP-AKA mechanism for non-3GPP device authentication. After the EAP-AKA procedure, the device is authenticated and further signaling (NAS) for registration proceeds. For transferring the NAS signaling a separate IP Sec SA is established. This will carry all further signaling messages between the device and 5G core. Further to this, the device can establish PDU sessions as required. The signaling flow for PDU establishment remains largely close to what we have seen in the 3GPP access case. Data traffic can proceed after the PDU is established. Further procedures such as service request for idle to connected mode transfer, etc. remain the same.

CHAPTER

8

Automation

8.1 Introduction

5G enables service providers to be an integral part of various businesses, driving growth by providing customized services beyond connectivity. This is very important for SPs constantly struggling with identifying new revenue opportunities.

At a high level there are significant additional complexities compared to the previous generations: it is access agnostic and has disaggregated radio access network (RAN); fronthaul and/or midhaul are introduced; the mobile core has new elements; the control and user planes are separated; and edge computing is introduced, to name a few. Understanding evolving requirements, stitching together the pieces of the new network, and transitioning services and traffic from a very different existing network can be a complicated task.

The 5G use cases and business models with the greatest potential for monetization currently revolve around automation and network slicing in vertical markets, such as manufacturing, transportation, and healthcare, which have some of the strictest and most diverse requirements for network performance. Current use cases indicate that private 5G networks, MEC platforms and hyperscale support are essentials in realizing any practical benefits from 5G upgrades. CSPs can use 5G, cloud and edge technologies to slice their private or shared networks and offer specialized services to meet the needs of vertical enterprises (Figure 8.1).

Figure 8.1: 5G automation.

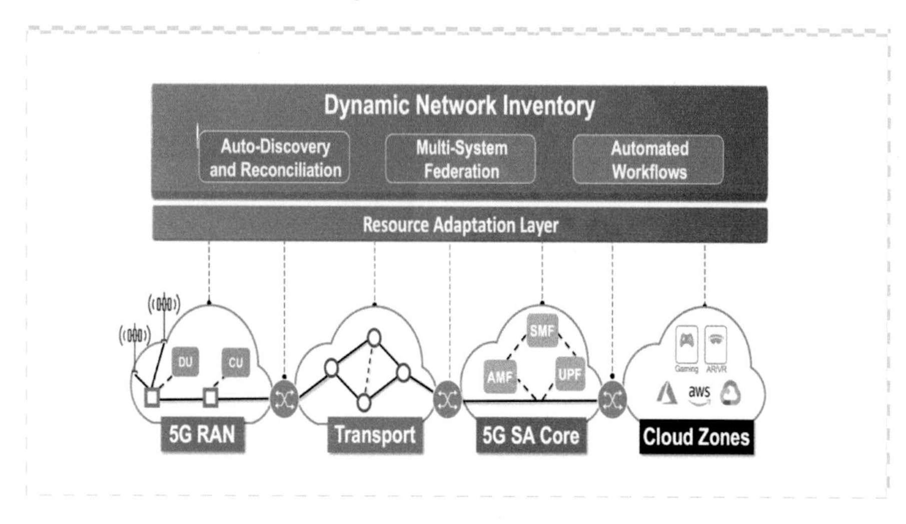

It must be cloud native. Container-based microservices create a loosely coupled architecture with modular components to break down and simplify the complexity of legacy frameworks. Along with open APIs and DevOps, cloud-native architecture provides near instantaneous scalability, deployment, and redeployment to any cloud platform for optimized flexibility, upgrades, and partner ecosystems.

It must react in real time. 5G networks will have to provide simultaneous responses and support for millions of devices running billions of applications and services. Each device, application and service will have different needs at different times. A network that cannot automatically and rapidly scale and adjust to accommodate these needs will defeat the entire purpose of automation by wasting a staggering number of resources.

It must support intent-based orchestration to align network automation with business needs amid the growing complexity and scale of 5G networks. Intent-based orchestration uses preconfigured service models, policy, and context with closed-loop control to automate the entire service and network slice life cycle. Closed-loop service management can establish effective intent-based orchestration; it uses AI/ML and analytics to monitor, manage and optimize everything from design to assurance for smooth operations.

5G network automation and orchestration is a technology that plays a critical role in managing the complexity of 5G networks. These networks are characterized by diverse services, multiple layers of hierarchy, and a vast array of technologies. Automation and orchestration, thus, become essential to efficiently operate and manage these networks.

5G network automation refers to the use of technology to perform network management tasks that would otherwise require human intervention. This includes tasks such as network configuration, fault management, and performance management among others. In essence, automation is about minimizing the manual effort needed to manage the network, which in turn reduces operational costs and improves network efficiency and reliability.

5G network orchestration, on the other hand, involves the coordinated management of different network resources to ensure they work together to deliver the required services. Orchestration includes processes such as service provisioning, network configuration, and resource allocation. By orchestrating the network resources, service providers can ensure that the 5G network delivers the expected performance, capacity, and coverage.

In the context of 5G, network automation and orchestration are essential for several reasons. First, they enable service providers to manage the complex 5G networks efficiently and effectively. Second, they facilitate the deployment and management of new services quickly, which is critical in the fast-paced 5G market. Third, they ensure that the network resources are used optimally, thereby improving the network performance, and reducing operational costs.

In conclusion, 5G network automation and orchestration are key enablers for the successful deployment and operation of 5G networks. They allow for the efficient management of complex networks, quick deployment of new services, and optimal use of network resources. As such, they are critical for the success of any 5G network.

8.2 5G Automation Demands

8.2.1 Slicing demand

Network slicing is a critical feature of 5G networks. It enables the creation of multiple virtual networks on a single physical infrastructure, with each slice being customized to meet the specific requirements of a particular service or application. This is essential in 5G, as it supports a wide range of services, from high-speed broadband to Internet of Things (IoT) applications, each with different performance requirements.

Automation plays a crucial role in the implementation and management of network slicing in 5G. Manual configuration and management of network slices could be labor-intensive and prone to errors given the complexity and dynamic nature of 5G networks. Therefore, automation is used to simplify these processes and make them more efficient.

For instance, automation can be used in the creation of network slices. It can automatically configure the network resources needed for each slice based on the specified service requirements. This can significantly reduce the time and effort needed to set up new services.

Automation is also essential in the management and operation of network slices. It can monitor the performance of each slice and automatically adjust the network resources as needed to ensure that the service level agreements (SLAs) are met. For example, if there is a surge in demand for a particular service, automation can quickly allocate more resources to the corresponding network slice to handle the increased traffic.

Moreover, automation can play a role in the maintenance and troubleshooting of network slices. It can detect and diagnose issues in real-time, and in some cases, automatically resolve them before they impact the service quality.

In conclusion, automation is crucial for the effective implementation and management of network slicing in 5G. It simplifies the configuration, operation, and maintenance of network slices, thereby improving the efficiency and reliability of 5G networks.

8.2.2 Necessity of automation

Single-pane-of-glass management: Centralized visibility of all deployed IP, optical and RAN devices across the distributed organization is key. A simplified workflow to deploy and update policies with few easy clicks/steps should be included. A hierarchical SDN solution should be able to automatically build and manage full mesh overlay topologies, underlay links, including LTE/5G wireless WAN options, for connectivity between sites. With guided workflows, automated overlay, and simplified business policies, IT staff hours spent on infrastructure deployment and changes are reduced from months to minutes.

Machine learning and AI: 5G technology, machine learning (ML) and artificial intelligence (AI) are increasingly intertwined, creating a powerful synergy that enhances the capabilities and user experience of the end-users. 5G networks service are offered with high data speeds, low latency, and massive scale device connectivity, generate an abundance of data that can be leveraged by ML and AI algorithms.

Machine learning and AI are used to harness the data generated within 5G networks to optimize network performance, automate operations, and provide intelligent services. For example, ML algorithms can predict network congestion and dynamically allocate resources to ensure consistent high-quality service.

AI-powered network management systems can perform real-time traffic analysis, enabling proactive problem resolution and enhancing security by identifying and mitigating threats swiftly.

Security: 5G security automation is a critical component of securing fifth generation (5G) wireless networks, as it leverages automated processes, machine learning, and artificial intelligence to continuously monitor, detect, and respond to security threats in real-time.

With the rapid expansion of 5G networks and the increasing number of connected devices, the attack surface for potential threats has grown significantly. Security automation is necessary to keep pace with the evolving threat landscape and protect the integrity, confidentiality, and availability of 5G services.

Some of the key aspects of 5G security automation include:

Threat detection and prevention: Automated security tools continuously monitor network traffic, devices, and applications to identify unusual or malicious behavior. Machine learning algorithms can recognize patterns associated with cyberattacks, and automated responses can be triggered to block or mitigate threats.

Anomaly detection: Security automation uses AI and ML to establish a baseline of normal network behavior. When anomalies are detected, such as unusual traffic patterns or unexpected device behavior, automated systems can flag potential security breaches for investigation.

Rapid incident response: In the event of a security incident, 5G security automation can respond swiftly. Automated actions may include isolating affected network segments, blocking malicious traffic, or deploying security patches or updates to vulnerable devices or network components.

Security policy enforcement: Automation ensures that security policies are consistently enforced across the 5G network. This helps prevent misconfigurations and policy violations that can lead to security vulnerabilities.

Scalability and efficiency: As 5G networks scale up to support a massive number of devices and applications, security automation is essential for efficiently managing and securing the network without the need for a proportionate increase in manual security personnel.

Open APIs: Open application programming interfaces (APIs) are essential for enabling third-party applications to interact with the 5G network, facilitating automation and innovation.

Policy-based control: Automation requires policy-driven control mechanisms that can adapt to changing network conditions and user requirements.

Multi-vendor capability: 5G orchestrators with multi-vendor capability play a pivotal role in ensuring the interoperability and seamless integration of diverse network components and services in a 5G environment. As 5G networks rely on a combination of hardware and software from various vendors, the ability to orchestrate and manage these heterogeneous elements is crucial for efficient network deployment and operation.

Key aspects of 5G orchestrators with multi-vendor capability include:

Vendor-agnostic approach: These orchestrators are designed to work with network equipment and software from different vendors, ensuring that network operators are not locked into a single vendor's ecosystem. This flexibility promotes healthy competition and allows operators to select the best components for their specific needs.

Interoperability: Multi-vendor orchestrators enable the integration of network functions, devices, and applications from different vendors, ensuring they work seamlessly together. This is particularly important in 5G networks, where various vendors may supply radio access equipment, core network functions, and other infrastructure components.

Vendor-neutral network slicing: Network slicing is a fundamental feature of 5G, allowing operators to create virtualized network segments tailored to specific services or use cases. Multi-vendor orchestrators facilitate the creation and management of network slices that can include components from different vendors, enabling operators to provide diverse and innovative services.

Reduced complexity: By offering a centralized management platform that spans multiple vendors, these orchestrators help reduce the complexity of network operations. They provide a single interface for network operators to configure, monitor, and optimize network elements, streamlining day-to-day tasks.

Enhanced flexibility: Multi-vendor orchestrators are essential for adapting to the evolving 5G landscape. They allow network operators to mix and match components as needed, facilitating network expansion, upgrades, and the introduction of new services without extensive manual intervention.

5G orchestrators with multi-vendor capability are instrumental in the efficient management of 5G networks, ensuring vendor diversity, interoperability, and adaptability while maintaining the high performance and quality of service that 5G promises.

Network analytics: 5G network analytics is a crucial component in the deployment and management of fifth generation (5G) wireless networks.

It involves the collection, processing, and analysis of vast amounts of data generated by the 5G infrastructure, including network nodes, devices, and applications.

This data driven approach provides network operators with valuable insights to optimize network performance, enhance user experiences, and ensure the efficient operation of the 5G ecosystem.

With 5G network analytics, operators can monitor network traffic in real-time, identify congestion points, and proactively address issues to maintain high-quality service. It also enables predictive maintenance, allowing operators to anticipate and resolve potential network problems before they impact users.

Additionally, 5G network analytics plays a pivotal role in the development of new 5G services, as it helps operators understand user behavior, preferences, and trends, allowing them to tailor offerings to meet evolving demands. In summary, 5G network analytics is an essential tool for managing and evolving 5G networks, ensuring their reliability, performance, and adaptability to meet the demands of modern communication and connectivity.

5G network analytics automation is a critical aspect of managing and optimizing the performance of fifth generation (5G) wireless networks. It involves the use of automated tools and machine learning techniques to collect, process, and analyze large volumes of network data in real-time, thereby enabling network operators to make data driven decisions for improved network performance and service quality.

Closed-loop automation in 5G networks is an advanced operational approach that uses real-time data and feedback mechanisms to continuously optimize and manage network performance. This process involves automated, self-correcting actions based on the monitoring of network conditions, performance, and user experience. It helps ensure that the 5G network operates efficiently, meets quality of service (QoS) requirements, and adapts to changing conditions automatically.

Key features of closed-loop automation in 5G networks include:

Real-time monitoring: The network constantly monitors various parameters such as traffic load, latency, and quality indicators. This data is collected from network elements, devices, and applications.

Analysis and decision making: Automation tools, including machine learning algorithms, analyze the collected data to identify anomalies or issues. These tools make decisions based on predefined policies and objectives.

Automated actions: If a problem or performance degradation is detected, the closed-loop system triggers automated actions to resolve the issue. These actions could include resource allocation adjustments, traffic rerouting, or scaling up or down network functions.

Continuous feedback: The system provides continuous feedback and performance measurements to assess the effectiveness of the automated actions. If the desired outcome is not achieved, the system can make further adjustments.

Dynamic optimization: Closed-loop automation is particularly valuable in dynamic 5G environments where network conditions and requirements can change rapidly. It ensures that the network adapts in real-time to maintain optimal performance.

Efficiency and self-healing: By automating corrective actions, closed-loop automation reduces the need for manual intervention, making network operations more efficient. It also contributes to self-healing capabilities, where the network can automatically recover from issues without human intervention.

Resource efficiency: Closed-loop automation optimizes resource utilization by dynamically allocating resources where they are needed the most, ensuring that network capacity is used effectively.

8.2.3 Challenges of moving to cloud

5G networks, including 5G core and edge, can be hosted in a public or private cloud, in a centralized or decentralized manner. A 5G network can be hosted entirely on a customer premise, or in an edge DC providing improvements in latency, availability, security, etc.

However, hosting the application in the cloud comes with its own challenges:

- New operations management and security solutions are required
- Finding use cases and business models behind the cloud edge
- Clouds must support the required high throughput
- Operations, processes, security, and availability must meet the expectations of SPs and their customers.

Cloud providers offer their own solutions to ease the design of moving services to the cloud. Disaggregation of functionality is typical in future networks. Some parts of the infrastructure (for example, a user plane function [UPF] or virtual centralized unit [vCU]) may need to be co-located with the application function (AF), and applications served from the cloud. With this new disaggregation comes additional parties to be involved, with corresponding clarifications about ownership, responsibilities, and management, that need to be made in a timely manner.

As 5G networks move towards cloud-based architectures, several challenges arise including network complexity, security concerns, and the need for seamless integration of services. Automation can play a key role in addressing these challenges.

Managing complexity: Cloud-based 5G networks involve a variety of technologies, protocols, and vendors, increasing their complexity. Automation can simplify the management of these networks. For example, it can automate the deployment and configuration of network services, reducing the time and effort required for these tasks.

Ensuring security: Security is a critical concern in cloud-based 5G networks. Automation can enhance security by enabling real-time monitoring of the network, detecting security threats, and responding to them promptly. It can also automate the implementation of security policies, ensuring they are consistently applied across the network.

Integrating services: In cloud-based 5G networks, services need to be seamlessly integrated to deliver a unified user experience. Automation can facilitate this by orchestrating the different network resources and services, ensuring they work together efficiently.

Scaling up and down: One of the main advantages of cloud-based networks is their scalability. Automation can make it easier to scale the network up or down based on demand. For instance, it can automatically allocate or de-allocate resources as needed, ensuring the network can adapt to changing traffic patterns.

Reducing operational costs: Automation can reduce the operational costs of cloud-based 5G networks. By automating routine tasks, it reduces the need for manual intervention, leading to cost savings.

In conclusion, automation can help solve the challenges associated with the shift to cloud-based 5G networks. By simplifying management, enhancing security, facilitating service integration, enabling scalability, and reducing costs, it can ensure these networks are efficient, reliable, and cost-effective.

8.3 Automation Architecture Components Overview

The automation solution architecture requires a product-agnostic approach to a 5G system architecture in a deployable operational framework. The solution shown below breaks down the various functions in the architecture to clearly define the necessary building blocks, while retaining the flexibility to work with the operational models and needs of all types of mobile service providers. The architecture of 5G network automation is typically composed of several key components, which work together to enable the efficient and effective management of the network. These components can vary depending on the specific design of the network, but generally include the following:

Figure 8.2 shows a high-level overall 5G automation architecture that can be divided into the following areas:

- A cross-domain layer that acts across multiple domains while also presenting the majority of customer-facing services.

Figure 8.2: 5G Automation architecture view.

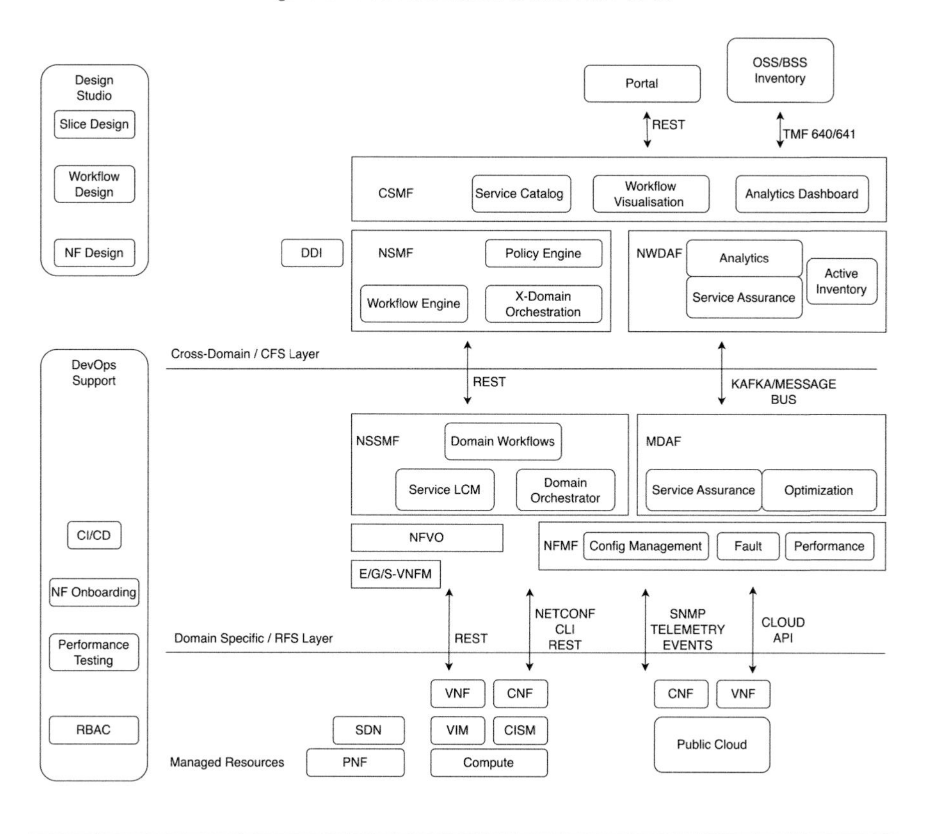

- Domain-specific layers that translate the intent of the cross-domain layer into the domain intelligence that manages the appropriate resource-facing services.
- Managed resources, which include virtualized, cloud-native, and physical network functions specific to a domain.
- DevOps support functions essential for automation and operation.

8.3.1 Customer centric service and resource centric service

The fundamental goals of a successful 5G automation architecture are to manage complexity and increase operational efficiency. A layered architecture that separates the specifics of the resource-facing services at the domain level from the customer-facing services at the cross-domain level is critical in fulfilling these goals. Key considerations for this separation are:

- Consistent abstractions
- Standardized interfaces
- Idempotent/reversible operations
- Clear distribution of control between domains.

The intent is to allow the resource-facing functions greater control of fine-grained operations at the domain level, while presenting abstracted operations for the customer-facing functions to compose slice-level operations from multiple domains.

8.3.1.1 Functions in CFS and RFS

The following functions are common in the RFS layer in all the domains:

NFVO and G-VNFM: Manage the lifecycle of virtualized or containerized workloads.

- NFMF: Manages configuration, fault, and performance of one or more individual network functions.
- NSSMF: Manages workflows and orchestrates and performs service lifecycle management at a domain level. It may also present some visualization capability such as topology.
- MDAF: Provides analytics and assurance capabilities at the domain level. Active inventory – the domain view of the state of all managed devices in order to provide correlation and running data – shall also be maintained while being synchronized with the overall static and physical inventory maintained by the OSS/BSS. It may also provide some domain-level optimization functions.

The following functions are present at the cross-domain/CFS layer:

- NSMF: Performs cross-domain slice management functions, utilizing a workflow engine and a cross-domain orchestrator. Performs policy enforcement and resource enforcement for all slice operations.
- NWDAF (network data analytics function): Provides analytics and assurance capabilities at the cross-domain (slice and service) level, while maintaining cross-domain active inventory. It will enable service-level, closed-loop operations via E-W integration with the NSMF.
- CSMF: Acts as a presentation layer – presents the service catalog, workflow visualization, and analytics/service assurance dashboard.

8.3.2 DevOps support functions

DevOps methodologies and their aligned support functions will be essential to deploy and operate the automation framework.

These functions are the glue that hold the framework together to provide the logistics and command-and-control functions that enable all disparate components and humans in the system to interact in a consistent, secure manner.

CI/CD: Supports the ability to rapidly test and deploy software and other network artifacts through the system.

NF onboarding: Supports the ability to rapidly add an NF or application to the service catalog.

AAA: Access, authorization, and accounting framework that the entire automation architecture will use – includes certificate servers, RBAC, LDAP integration, audit mechanisms, and SOC integration.

8.4 Domain specific Orchestration

8.4.1 RAN orchestrator

ITU R-M.2083 outlines the eMBB, URLLC, and mMTC use cases for the RAN in the context of 5G. The implementation of these use cases demands a substantial level of automation on a scale that poses challenges. Moreover, the 5G paradigm introduces crucial concepts such as disaggregation into edge data centers and the virtualization of the RAN.

In this scenario, the network slice selection management function (NSSMF) plays a pivotal role by supporting both virtualized and containerized workloads through the network function virtualization orchestrator (NFVO) and virtual network function manager (VNFM). Simultaneously, it manages certain physical network functions (PNFs), including cell site routers, microwave radio links, cell site switches, and radio interface units (RIUs).

The architectural framework shown in Figure 8.3 aligns seamlessly with O-RAN specifications, particularly for non-real-time control and various service provisioning activities.

In this practical scenario, relevant to the implementation of 5G in a virtualized setting, the activation of a new cell site involves the automatic configuration of the radio interface unit (RIU), virtualized distributed unit (vDU), and virtualized central unit (vCU). Beyond the site initialization process, the RAN network slice selection management function (NSSMF) must handle

Figure 8.3: G-RAN orchestration.

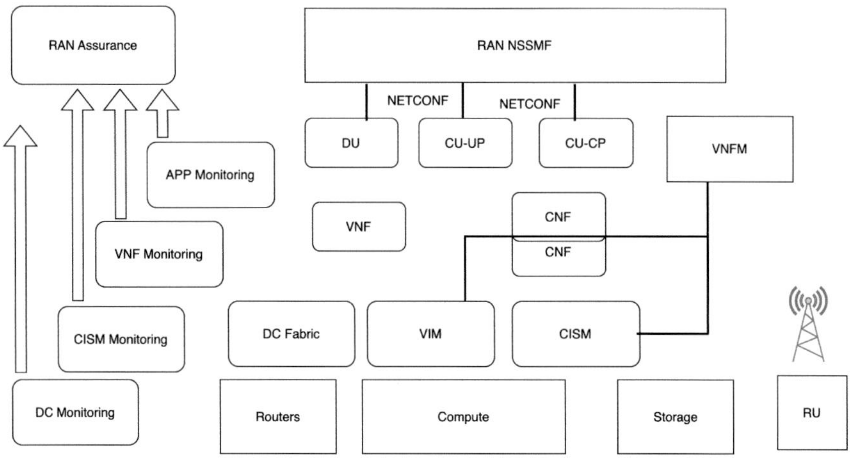

specific slice scheduling configurations to meet the stringent requirements of each distinct slice.

To establish a new slice, the NSSMF executes the following steps:

- Upon receiving a NSSI instantiation request, preliminary checks are conducted to ensure that the current NSSI can accommodate the new request effectively.
- The availability of RAN resources is examined to verify their capacity for provisioning the required scheduling rules for the new slice.
- The NSSMF configures the essential scheduling information for each RAN slice based on existing templates (NSMT). This meticulous process ensures the adherence to the unique requirements of each slice within the network. The configuration requirements are sent to RAN NFMF for configuration management and further for service assurance.
- The configuration needs are forwarded to the RAN NFMF for the purposes of managing configurations and subsequently ensuring service assurance.

8.4.2 Transport orchestrator

The transport infrastructure plays a crucial role in supporting 5G services across various network domains (Figure 8.4). Achieving a seamless end-to-end service lifecycle underscores the need for establishing a unified transport

Figure 8.4: Transport orchestration.

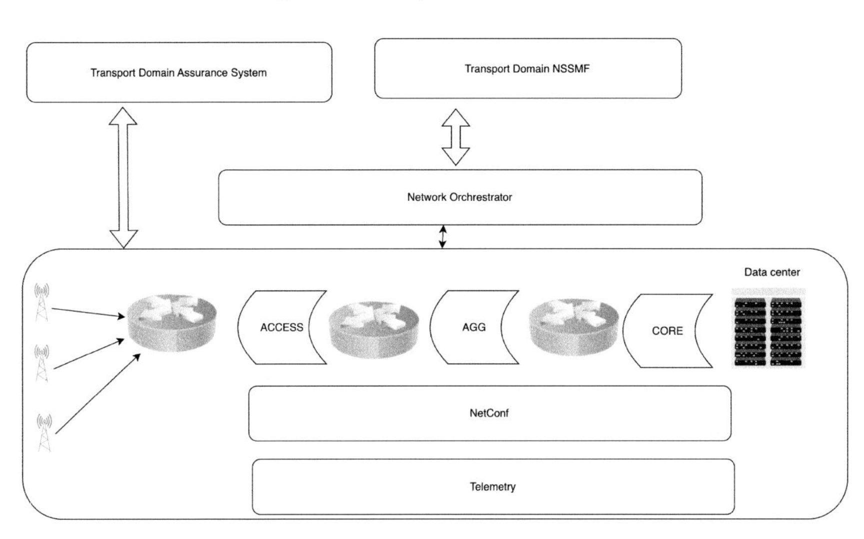

architecture. From a broad perspective, the mobility transport network typically encompasses xHaul and core domains.

Both network domains, incorporating routers that may be virtualized or physical based on their functions, actively participate in the implementation of network slicing. By harnessing the advanced programmability and automation features of transport software-defined networking (T-SDN) atop a unified transport architecture built on segment routing (SR), the need for frequent device-level configuration changes related to the introduction or removal of new slices can be significantly reduced or, in certain cases, eliminated.

Before delving into the specifics of network slicing for transport, it is imperative to examine the impact of 5G on the xHaul architecture. The introduction of the RAN split concept transforms xHaul to encompass fronthaul, midhaul, and backhaul networks.

The RAN split imposes stringent requirements on xHaul, demanding high bandwidth, low latency, and robust availability. These specifications are crucial to support precise clock synchronization among disaggregated radio functions, accommodate additional capacity due to radio densification, and facilitate mmWave deployment.

Depending on the placement decision of the user plane function (UPF), the backhaul transport network (N3 interface) may be minimal (UPF closer to the RAN edge) or extensive (UPF placed centrally or even further, especially for certain non-latency enterprise application use cases).

Beyond the three service types (eMBB, URLLC, and mMTC), network slicing within the network transport infrastructure must fulfill additional requirements:

Transport slice management: The capability to create, modify, and delete a 3GPP network slice, including necessary actions at the transport layer. Monitoring the health and performance of the slice through operations, administration, and maintenance (OAM) capabilities is essential.

Slice isolation: Ensuring each transport slice is isolated from others to meet stringent SLAs. Independent slices must adhere to proper quality of service (QoS), performance, security, operational, and reliability levels.

Resource reservation: The ability to reserve transport resources for a specific transport slice to meet QoS requirements.

Abstraction: Utilizing resources required to model and build a transport infrastructure to meet the demands of a network slice.

Network slicing in terms of hard and soft types must meet specific requirements, with hard slices having dedicated resources, while soft slicing allows resource sharing with the ability to return resources to the network when not in use.

The transport layer, responsible for connecting all other domains for all customers, demands continuous resource optimization and traffic management, requiring varying levels of network automation depending on the use case. Offline planning involves simulating scenarios without affecting the working network, while online optimization continuously monitors network conditions and SLA measurements, redirecting flows using AI-driven scenarios.

This setup necessitates a standards-based and vendor-agnostic approach to provide a unified domain management solution for a potentially multi-vendor environment within a domain. Service stitching between different domains requires the cross-domain orchestrator to manage overlays, such as MPLS-SR to VXLAN, or maintain an end-to-end MPLS-SR or SRv6 topology across diverse domains.

8.4.3 DC orchestrator

The infrastructure deployed at various levels of DC placement, such as central, near-edge, and far-edge, plays a pivotal role in the orchestration and automation of 5G services. This encompasses the lower segment of the ETSI MANO stack, considering both physical components (compute, storage, network) and the virtualization layer, facilitated through a virtualized infrastructure manager (VIM) or cloud orchestration.

A critical consideration involves the distribution of data sources and applications. Low-latency applications necessitate further distribution to the edge, aligning with end-users and their capabilities. Avoiding constant backhaul or traffic distribution from centralized data centers proves beneficial for various applications and services.

The introduction of edge computing, specifically multi-access edge computing (MEC), based on a distributed data center architecture, not only enhances application interactions with end-users but also unlocks new possibilities for applications and services.

While managing a centralized data center deployment presents its own set of challenges, proactively handling infrastructure, applications, data sources, and workloads distributed across multiple locations elevates complexity to a new level (Figure 8.5). To effectively navigate this complexity and usher in a new

Figure 8.5: Data center orchestration.

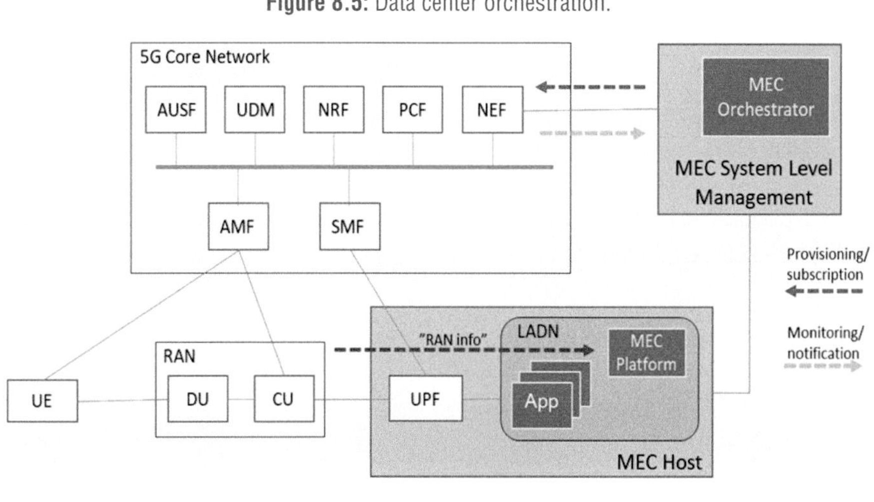

era of 5G-enabled services, an automation and orchestration solution becomes crucial. This solution should encompass key aspects like network fabric management, including service chaining functions and resource allocation per network slice; network service lifecycle management, covering instantiation, scaling, updating, and termination of services; VNF lifecycle management, involving instantiation, scaling, updating, and termination of VNFs; and NFVI resources supporting virtualized and partially virtualized network functions through abstracted services for network, compute, and storage.

8.4.4 Packet core orchestrator

The 5G mobile core comprises several novel network functions (NFs), and their setup, capacity, and dynamics, influenced by the use case, vary across elements such as AMF, SMF, UPF, PCF, AUSF, UDM/UDR, NSSF, NRF, and NEF.

Most, if not all, of these NFs adopt a cloud-native architecture, characterized by small, independent, and loosely coupled services. This shift is significant as it introduces changes and challenges in terms of deployment and service assurance compared to previous generations.

In the realm of 5G core automation, including potential deployment at edge locations, a notable infrastructure challenge is the coexistence of virtualized and cloud-native network functions (VNFs and CNFs), operating on different platforms yet requiring consistent management and orchestration. This coexistence is anticipated in data centers (DCs) in the foreseeable future, necessitating automation solutions that address instantiation, service deployments, and management for both types.

All elements of domain automation and orchestration form part of the resource function steering (RFS) layer in the architecture. They seamlessly integrate with cross-function steering (CFS) layer elements like the cross-domain data collector, network slice selection management function (NSMF), and network function management function (NFMF) through protocols such as REST, NETCONF, SNMP, or CLI.

As per ETSI standards, integration between the network function virtualization orchestrator (NFVO) and NSSMF, both situated in the RFS layer, is expected to occur through the SOL005 interface, particularly when NFVO and NSSMF are distinct network functions.

Key requirements for the 5G core domain orchestrator involve prechecking the necessity to instantiate new NFs for a new service or network slice subnet instance (NSSI), applying a predefined network slice management template

(NSMT) to meet service requirements, instantiating necessary 5G core NFs through integration with existing NFVO using the SOL005 interface, integrating with NFMF for NF service configuration (Day 1, 2, and N), and collecting telemetry from NFs and potentially from the cloud infrastructure for service assurance.

Emphasizing the preference for a vendor-agnostic approach in the NSSMF and NFMF setup is crucial to simplify the overall automation solution, given the expected multivendor environment in the 5G core network. A vendor-agnostic 5G core domain orchestrator should support flexible workflow management and offer a broad spectrum of configuration, fault, and performance management options.

8.5 Cross Domain Orchestration

To realize the objectives of 5G automation, it is imperative to implement cross-domain orchestration that links various components across different network domains. Positioned within the integration layer of the solution, the cross-domain orchestrator has the capability to receive service instantiation requests from the operations support system/business support system (OSS/BSS) or a self-service portal. It actively engages in fulfilling these service requests and performs additional functions. The cross-domain orchestrator is anticipated to interact with distinct domain orchestrators situated in the domain-specific layer to successfully execute the assigned tasks.

The cross-domain orchestrator plays a vital role in end-to-end (E2E) service delivery and E2E closed-loop automation, contributing various essential functions:

Integration with OSS/BSS layer: It interfaces with the OSS/BSS layer, receiving order management requests (TMF 640/641) and exporting available services on the service catalog to the BSS digital marketplace (TMF 633), particularly when TMF interfaces are not yet ready on OSS/BSS, and REST API integration is offered.

Service fulfilment module: This module translates order requests into fulfillment logic, encompassing pre-checks, service breakdowns, policies, and post-checks.

Integration with service layer and domain orchestrators: It integrates with the service layer and potentially different domain orchestrators to fulfill requests at the resource level.

Service inventory: It maintains a comprehensive list of all instantiated services.

Service topology view: It provides a visual representation of the service topology.

Service design studio: This feature allows the creation of service logic using a graphical user interface (GUI).

Service analytics module: It receives telemetry and performance information from various resource function steering (RFS) layers, consolidating key performance indicator (KPI) management, conducting advanced analytics, and applying logic for closed-loop automation.

Closed-loop automation for service remediation: The orchestrator incorporates closed-loop automation for the remediation of services.

It's important to note that the self-service portal in this solution does not replace the self-service portal on the BSS layer (digital marketplace). The latter provides integration with the billing system, customer relationship management (CRM), or any other necessary business support system.

Several integrations are needed to fulfil the cross-domain orchestration functionality (Figure 8.6).

For effective communication, an interface is necessary to interact with various components, including service activation and configuration, service catalog, and service inventory. This interface can be established using either general REST APIs when connecting with the business support system (BSS) or marketplace, or it can leverage TMF standard APIs. The TMF standard APIs, specifically TMF640/641 Service Activation and Configuration API [9][10], TMF638 Service Inventory API [11], and TMF633 Service Catalogue API [12], provide standardized protocols for seamless communication and integration within the telecommunications ecosystem. These APIs play a crucial role in streamlining operations and facilitating the exchange of information related to service activation, configuration, inventory management, and catalog representation.

A communication interface is essential for interacting with both service activation and configuration, as well as the service catalog and service inventory. When interfacing with the business support system (BSS) or marketplace,

Figure 8.6: Cross-domain integration.

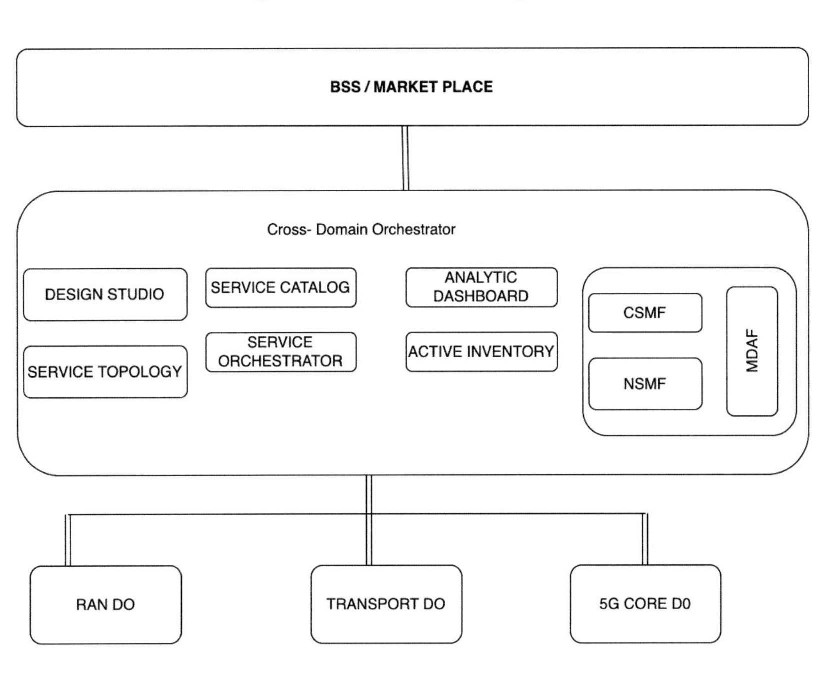

general REST APIs or standardized TMF APIs can be employed. These TMF standard APIs include:

- TMF640/641 Service Activation and Configuration API
- TMF638 Service Inventory API
- TMF633 Service Catalogue API.

For southbound integration, connecting with various domain controllers typically involves the use of NETCONF or REST APIs. In cases requiring direct integration with the NFVO, the SOL005 interface is employed. It's noteworthy that the 3rd Generation Partnership Project (3GPP) is currently in the process of defining specific standard APIs for integration with different RAN NSSMF and 5G core NSSMF in releases 16 and 17.

8.6 Service Assurance and Operations

The 5G network platform, incorporating advancements in radio technology, cloud core infrastructure, and the software-defined transport layer, is establishing a more dynamic networking environment. This dynamic nature necessitates automated, closed-loop network optimization to uphold service level agreements (SLAs) on a per-service and network-slice-by-network-slice basis. The role of 5G service assurance becomes crucial in ensuring quality of service (QoS), quality of experience (QoE), and SLAs across hybrid physical, virtual, and cloud-native networks. These networks encompass services spanning radio access network (RAN), transport, packet core, data center (DC), and application domains. Effective data driven analytics are indispensable for the dynamic provisioning, scaling, and termination of network slicing. Network slicing allows diverse customers to have distinct experiences with specific QoS, QoE, and SLA parameters managed on an individual slice basis.

The correlation of fault and performance key performance indicators (KPIs) with service quality experience key quality indicators (KQIs) and underlying resources is crucial to ensure per-slice SLA adherence and a balanced subscriber experience with resource consumption. Given the vast data volumes and the dynamically distributed nature of the network, artificial intelligence (AI) and automation are essential to enable the dynamic creation, scaling, and termination of 5G network slices.

Centralized assurance, monitoring, and analytics are essential components of network management and optimization, especially in the context of complex and high-performance networks like 5G (Figure 8.7). These functions involve the collection, analysis, and visualization of network data and performance metrics from various sources in a centralized location, allowing network operators to gain insights, make informed decisions, and ensure the efficient operation of the network.

Data collection: A centralized assurance system collects data from various network elements, including radio access, core network, and edge computing components, as well as user devices and applications. This data includes performance metrics, traffic patterns, device behavior, and security events.

Real-time monitoring: Centralized monitoring tools continuously track the state of the network, device performance, and application behavior in real-time. This allows operators to detect issues, bottlenecks, or anomalies as they occur and take proactive measures to maintain network performance.

Data analysis: Centralized analytics platforms use advanced algorithms, including machine learning and artificial intelligence, to analyze the collected data. This analysis can reveal trends, correlations, and potential problems, such as traffic congestion, security threats, or performance degradation.

Figure 8.7: 5G service assurance.

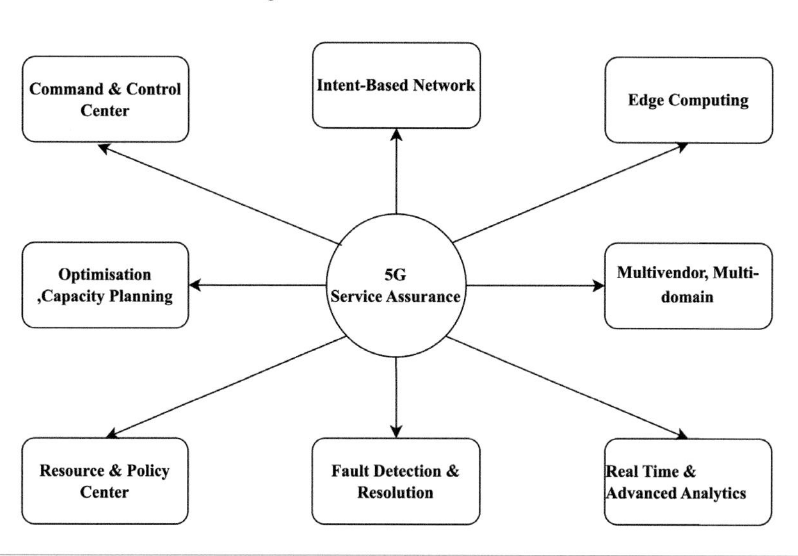

Visualization and reporting: The insights gained from data analysis are presented in the form of visual dashboards, reports, and alerts that provide network operators with a clear and actionable view of network health and performance.

Performance optimization: Based on the analytics results, operators can make informed decisions to optimize network performance, allocate resources more efficiently, and implement changes or upgrades to meet user demands.

Security monitoring: Centralized assurance systems also monitor security events and threats across the network. They can detect and respond to anomalies, intrusions, and vulnerabilities in real-time to enhance network security.

Resource allocation: With centralized analytics, network operators can dynamically allocate network resources where they are needed most, optimizing capacity utilization and minimizing latency.

8.6.1 End–end automation and assurance solution

Creating a schematic for an end-to-end automation and assurance solution can be a complex task, as it may involve numerous components and systems.

However, it can provide a simplified outline of the key elements you might find in such a solution:

User interface: The solution typically starts with a user interface or dashboard where users (e.g., network administrators or operators) interact with the system. This is where they initiate and monitor automation and assurance processes.

Order management: The system may include an order management component where users submit requests for various services or products. These orders can be for network provisioning, service activation, or other tasks.

Automation orchestrator: The heart of the solution is the automation orchestrator. This component manages the automation workflows, ensuring that tasks are executed in the correct sequence and in compliance with predefined policies and standards.

Inventory management: An inventory management system keeps track of all network and IT assets, such as hardware, software, and licenses. It plays a vital role in provisioning and maintaining resources.

Configuration management: Configuration management tools store and manage configuration templates for different services and products. These templates guide the automated provisioning and configuration processes.

Resource allocation: The resource allocation component determines the allocation of computing, storage, and network resources based on the requirements specified in the orders.

Provisioning engine: This engine is responsible for the actual provisioning of products or services. It configures hardware and software components and deploys the necessary resources.

Monitoring and assurance: Continuous monitoring is crucial for ensuring the quality and performance of provisioned services. This component collects performance data, logs, and alerts from the network, applications, and devices.

Analytics and reporting: Analytics tools process the monitoring data, enabling the system to make real-time decisions and offering insights into network and service performance. Automated reports are generated for stakeholders.

Security and compliance: Security tools ensure that provisioning and configuration changes adhere to security policies and standards. They also detect and respond to security threats.

User notifications: Automated notifications are sent to users and customers, informing them of the successful provisioning of services, access credentials, and any relevant information.

Billing and invoicing: If applicable, a billing and invoicing module integrates with the solution, automatically generating invoices based on the provisioned services and usage.

Scaling and optimization: This component manages the dynamic allocation and scaling of resources to optimize performance and cost-efficiency based on real-time data.

Feedback loop: The solution may include a feedback loop that provides continuous improvement by learning from past automation and assurance processes.

Integration and APIs: The solution should be able to integrate with various network devices, software, and systems through APIs, enabling seamless communication and data exchange.

Lifecycle management: Automated provisioning is only one part of the solution. The system should also support the ongoing lifecycle management of services, including updates, scaling, and decommissioning.

Please note that the specific components and their complexity can vary significantly depending on the use case and industry. A comprehensive end-to-end automation and assurance solution should be designed to meet the unique requirements of the organization and the technologies it employs.

CHAPTER

9

Network Slicing

9.1 Why is Network Slicing Important?

Service providers require a method and idea to efficiently reuse and oversee their capacity, and this is where the concept of network slicing comes into play. The noteworthy aspect of network slicing lies in its attainability through the swift advancement of SR and SRv6 solutions. Network slicing is also one of the requirements in 5G.

The term "Network slicing" in Ethernet networks has been around since the inception of virtual local area networks (VLANs). The concept has gained greater prominence with the emergence of software-defined networking (SDN) and, more recently, software-defined wide area networks (SD-WANs), which extend SDN principles to wide area networking.

Currently, the business needs of the majority of service provider networks include the following:

- Offering a greater variety of services.
- Facing limitations in expanding capacity and deploying service-specific nodes in the service provider network due to potential increases in capital and operational expenditures.
- Recognizing that not all services experience the same load levels throughout the day or week.
- Emphasizing the importance of network reusability when application loads vary, leading to enhanced flexibility and reliability.

9.2 Network Slicing in 5G

In the era of 5G, the traditional methods of orchestrating and ensuring services, characterized by siloed approaches and technology-centric tools and processes, are no longer sufficient. To support digital services and network slicing efficiently and expansively, the service provider's operations support system (OSS) must evolve to unlock the network's value across diverse domains, technologies, and vendors.

This transformation requires transitioning from a focus on technology to a more business-centered approach, where automation is aligned with business goals and defined service level agreements (SLAs). Within the framework of established 5G architectures, service providers aspire to assign dedicated portions of their network to cater to the distinct requirements of customers. This flexibility enables scalable modifications to services. Illustrative scenarios (Figure 9.1) encompass supporting the Internet of Things (IoT) in manufacturing, overseeing autonomous vehicles within a transportation fleet, or separating AI-driven video analytics from point-of-sale information in a retail environment.

Figure 9.1: High level view of slicing in 5G.

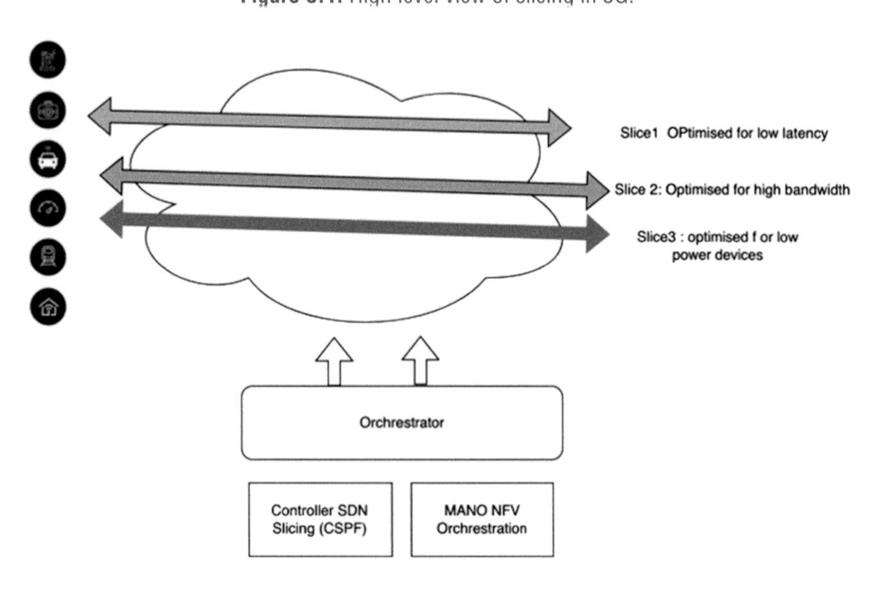

To illustrate further, a 5G network slice could specifically support autonomous forklifts in a factory. This support ensures seamless communication, safeguarding nearby factory workers even during spikes in communication traffic from other areas within the factory.

As described in the previous chapters, one of the strengths that is underlined by 5G mobile networks is the ability to provide broadband and low latency connections, even when you have many objects connected to the network and differentiated service as per application.

Every use case comes with its specific demands, and quite a few of these requirements can be independent in terms of quality, facilitating efficiency in scenarios where infrastructure is shared.

For instance, in the context of an autonomous vehicle application, which necessitates low latency and high reliability for safety, as opposed to a virtual reality use case, where high throughput is essential with somewhat relaxed reliability since a few lost pixels, frames, or lower resolution might be acceptable. Those two quality parameters could be granted, at the same time, by the same physical infrastructure. On top of the same hardware a network slice can be built to support the vehicle network, with specific KPIs in terms of latency, and another network slice to provide large throughput for VR applications. Some use cases are shown in Figure 9.2.

Figure 9.2: Different slicing use cases under major 5G use cases.

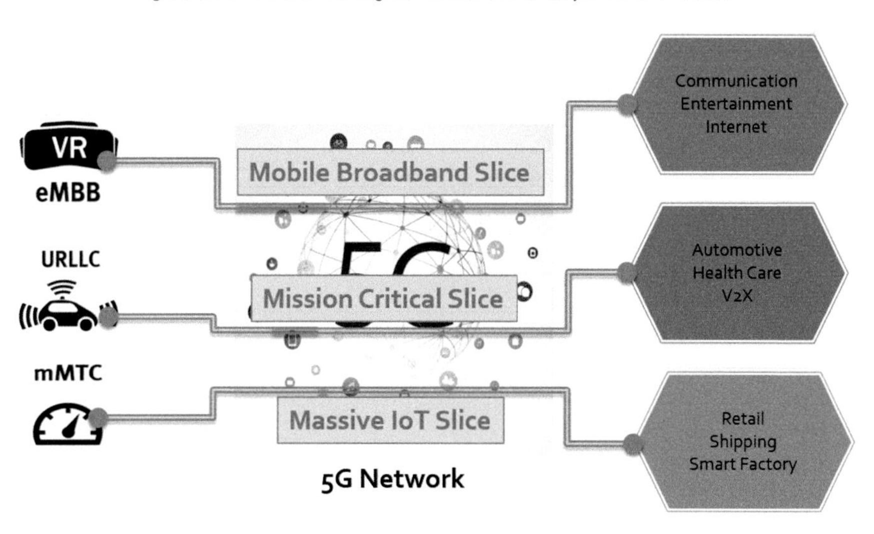

9.2.1 Cross-discipline SDN and network slicing

The implementation of NFV/SDN (network function virtualization/software defined network) in 5G networks is currently a reality and is anticipated to experience significant growth in the coming years. This evolution will facilitate network slicing, enabling the creation of highly adaptable network instances tailored to the specific requirements of applications, services, and operator business models.

Both NGMN and 3GPP have been developing the definition and use cases for network slicing so that the SDOs (standards development organizations) can provide detailed studies to understand the features and functionalities that will be required for network slicing.

5G network slicing is a network architecture that enables the multiplexing of virtualized and independent logical networks on the same physical network infrastructure. Each network slice is an isolated end-to-end network tailored to fulfil diverse requirements requested by a particular application.

For this reason, this technology assumes a central role to support 5G mobile networks that are designed to efficiently embrace a plethora of services with very different service level requirements (SLR). The realization of this service-oriented view of the network leverages on the concepts of software-defined networking (SDN) and network function virtualization (NFV) that allow the implementation of flexible and scalable network slices on top of a common network infrastructure.

From a business model perspective, each network slice could be administered by a mobile virtual network operator (MVNO). The infrastructure provider (the owner of the telecommunication infrastructure) leases its physical resources to the MVNOs that share the underlying physical network. According to the availability of the assigned resources, a MVNO can autonomously deploy multiple network slices that are customized to the various applications provided to its own users.

9.2.2 5G network slicing deployment challenges

SPs will need to ensure that applications are aligned to the right service level – and therefore network slice – and measure the actual performance to see that it delivers.

Another potential challenge lies in deciding how many slices are adequate for customers' service needs. Provisioning network slices with excess capacity or insufficient resources could lead to unnecessary complications in overseeing a customer's entire network and operations.

9.3 Slicing Characteristics

Network slicing is a type of functionality that enables multiple independent networks to exist on the same physical network, using different "slices" of the same spectrum band. This allows organizations to accommodate different application requirements for security, reliability, and performance on the same network (Figure 9.3).

Figure 9.3: Slicing characteristics.

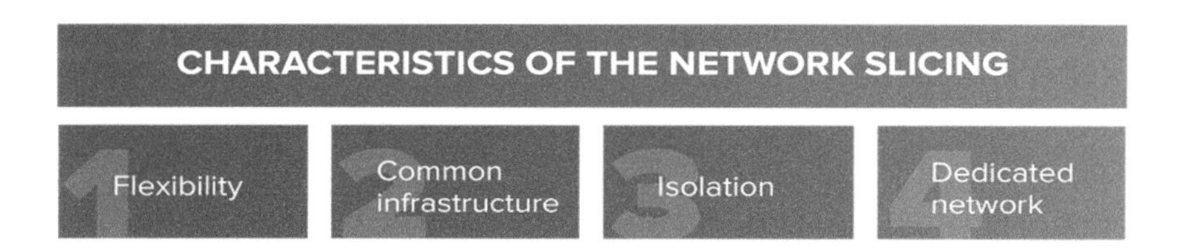

The implementation of network slicing spans end-to-end, covering the core to the radio access network (RAN). Within the RAN, slicing can be constructed based on physical radio resources, including transmission point, spectrum, and time, or on logical resources derived from the abstraction of physical radio resources. The deployment of network slicing takes advantage of emerging underlay transport technologies such as SR, SRv6, SDN, NFV, and automation, allowing for the rapid segmentation of the network and its resources to support specific applications, devices, domains, and groups. Utilizing network slicing proves to be a cost-effective approach for enterprises to fulfill service level agreements (SLAs) and ensure continuous allocation of the necessary resources for each application.

Slicing creates software-based, logical partitions within a self-contained, virtualized environment that takes precedence over physical components. Subsequently, any available capacity in shared resources, such as storage and processors, can be redirected based on business requirements.

The implementation of network slicing empowers a service provider (SP) to foster innovation by segregating and prioritizing mission-critical services above those with lower performance sensitivity. This approach allows the creation of multiple logical networks on a common shared physical network, effectively segmenting different parts of the network for various users and/or use cases. The capability to slice networks is more readily achievable in 4G and 5G, making this feature more realizable.

9.4 Slicing Domains

Slicing can be applied under the following domains of the network:

- RAN
- Core
- Transport
- Service.

Figure 9.4: Slicing layers and domains.

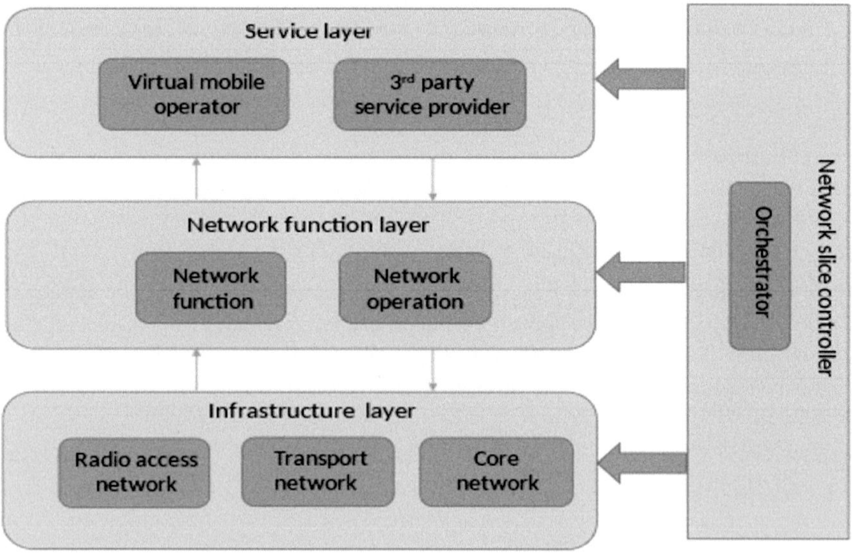

9.4.1 RAN slicing

RAN slicing is dynamic radio resource allocation and prioritization for different slices to fulfill the slicing requirements of different use cases. It involves partitioning the radio access network into separate and isolated segments to cater to specific service requirements, applications, or customer needs.

RAN slicing involves the creation of logically isolated segments within the radio access network. Each slice is dedicated to specific use cases, services, or applications. Each RAN slice operates independently, providing isolation from other slices. This isolation ensures that the resources and performance characteristics of one slice do not impact another.

RAN slicing allows for the customization of network parameters such as latency, throughput, and connection density to meet the specific requirements of different applications or services. Different slices can be tailored to support specific services or applications. For example, a slice may be optimized for enhanced mobile broadband (eMBB) services, while another may prioritize low-latency communication for critical IoT applications.

RAN slicing enables efficient allocation of resources such as bandwidth, frequency spectrum, and processing power. This ensures that each slice receives the necessary resources to meet its performance and quality of service (QoS) requirements. It is designed to be dynamic and adaptable. Network operators can adjust slice configurations in real-time to respond to changing demands, allowing for optimal resource utilization.

RAN slicing is often integrated with core network slicing to provide an end-to-end slicing experience. This integration ensures seamless connectivity and coordination between the RAN and the core network. RAN slicing supports a wide range of use cases, including enhanced mobile broadband, massive machine-type communication (mMTC), and ultra-reliable low-latency communication (URLLC). It is particularly valuable for industries such as healthcare, manufacturing, and autonomous vehicles that have diverse and stringent connectivity requirements.

9.4.1.1 5G RAN slicing concepts

1. Spectrum sharing and slice isolation

Implementing slicing in the RAN poses a significant challenge, particularly in efficiently designing and managing multiple slices on a shared infrastructure while ensuring the agreed service level agreements (SLAs) for each slice. This leads to the concept of "slice isolation," which aims to prevent any mutual impact between coexisting slices. A widely accepted understanding is that two slices are considered isolated if the actions taken on one slice do not lead to a violation of SLA on the other slice.

2. Slice identification

It is important to identify the slice to which a packet flow belongs, enabling a RAN network function to potentially apply specific treatment. Consider the scheduler, which currently manages a limited number of traffic classes, exemplified by the QoS class identifier (QCI). However, with the introduction of slices each having distinct service level agreements (SLAs), the task becomes more challenging for these network functions.

To address this challenge, a RAN slice management entity could simplify the process by:

- Executing function chaining for flows based on the SLAs associated with their respective slices.
- Choosing the appropriate configurations at the PHY layer for packets from flows belonging to a particular slice.
- Mapping the slice SLA to a QoS identifier and a tile to streamline scheduler complexity.

This approach facilitates the efficient handling of packet flows within diverse slices, ensuring that network functions can implement tailored treatments based on the specific requirements of each slice's SLA.

3. Tiling, scheduling, and puncturing

The principle is that time-frequency resources with the same numerology are grouped together within a tile (or radio block group). This reduces the processing burden associated to scheduler.

The scheduler plays a crucial role in resource allocation to meet service level agreements (SLAs) with diverse quality of service (QoS) requirements for various slices, each potentially having highly varied demands. To streamline the complexity of the scheduler, two approaches are considered: (i) dynamically determining the composition of tiles and (ii) mapping the slices to these tiles. This mapping facilitates a more straightforward management of numerology and other physical/MAC parameters, including transmission time interval (TTI) length.

It's worth noting that the scheduling complexity is alleviated by leveraging this mapping strategy, allowing for easier handling of diverse requirements among slices. Regarding time multiplexing, the 3GPP standard ensures symbol alignment between tiles, ensuring orthogonality. For frequency multiplexing, 3GPP recommends the insertion of guard bands between tiles with different subcarrier spacings to maintain orthogonality between adjacent tiles.

For URLLC services, the concept of puncturing is important to preempt resources as the traffic could be bursty in nature. In both downlink and uplink, for example, an eMBB transmission can be preempted to accommodate URLLC packets. eMBB packets could recover later by retransmissions.

9.4.1.2 Slicing in O-RAN

Network slicing stands out as a crucial application for O-RAN, given its inherent virtualization and built-in intelligence. O-RAN Alliance specifications govern the management of resources within a slice and the optimization of resources across different slices. The primary challenge lies in safeguarding the isolation of resources within a slice, ensuring that the resources utilized by one slice remain protected from other slices. Additionally, it is essential to dynamically scale resources in and out to guarantee compliance with service level agreements (SLAs) within a specific slice.

Figure 9.5 illustrates the O-RAN slicing architecture, encompassing slice management functions and various O-RAN architectural components. O-RAN adheres closely to 3GPP architecture principles, incorporating 3GPP-defined NSMF and NSSMF, extended to accommodate O-RAN network functions.

Within the O-RAN slicing architecture, the non-RT RIC (RAN intelligent controller) accumulates long-term data related to slices by interacting with the SMO. The near-RT RIC facilitates near-real-time optimization of RAN slice subnets through the execution of slicing-related xApps, communicating essential parameters to O-CU and O-DU via the E2 interface.

The O-CU, comprising a singular O-CU-CP and potentially multiple O-CU-UP(s) communicating through the E1 interface, is required to support slicing. The O-CU stacks, representing upper-layer protocols of the RAN stack, must exhibit awareness of slices and execute strategies for slice-specific resource allocation and isolation. Similarly, the O-DU, responsible for running lower-layer protocols of the RAN stack, should support slice-specific resource allocation strategies.

In response to performance monitoring (PM) requests from SMO and near-RT RIC, the O-CU/O-DU generates and transmits specific PMs through O1 and E2 interfaces, respectively. These PMs serve the purpose of slice performance monitoring and ensuring adherence to slice service level agreements (SLAs).

Figure 9.5: O-RAN slicing architecture.

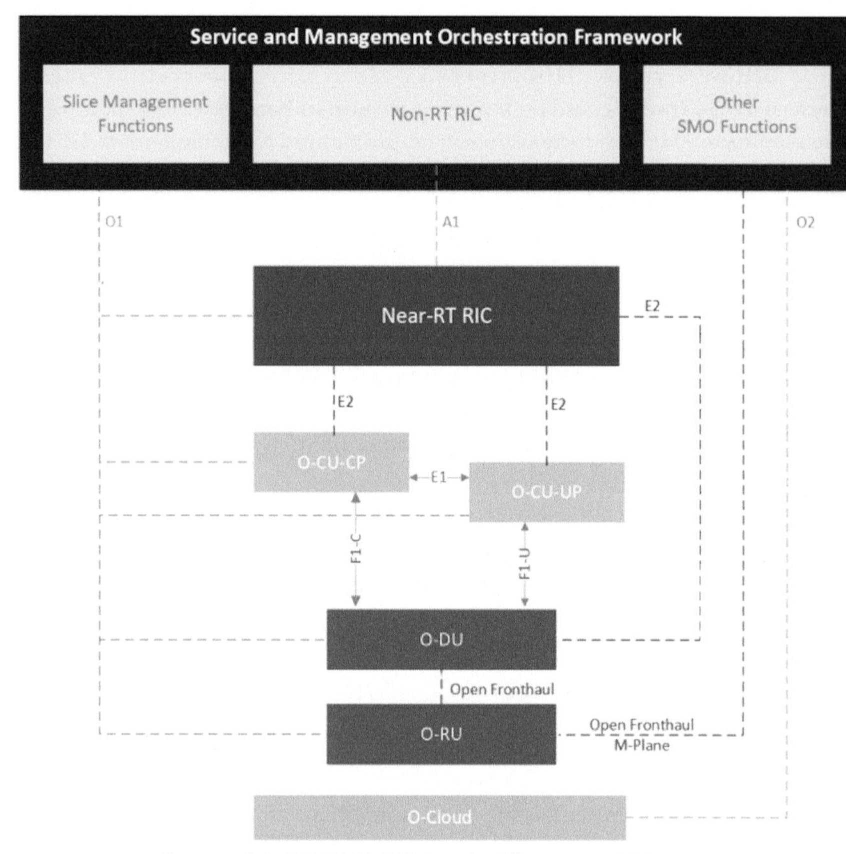

Source: O-RAN.WG1.Slicing-Architecture-R003-v11.00

9.4.2 Core slicing

5G network functions are containerized to support deployment of such functions more granularly than with monolithic designs. This enables rapid instantiation and the ability to scale only those functions that need to be scaled. However, this calls for an orchestration capability that performs more than just instantiating the CNFs, but also to manage them.

The 5G core orchestrator should inter-work with the cross-domain slice orchestrator and facilitate the lifecycle operations of a core network slice instance, including slice instantiation, provisioning, service chaining, modification, and termination.

The mobile core network functions designed for 5G standalone (5G SA) will be implemented as cloud-native functions (CNFs) utilizing cloud-native computing technologies like Kubernetes (K8s). A cloud-native architecture is not only a necessity for 5G networks but also for the applications deployed on these networks. Service providers must adopt this architecture, characterized by containerized applications built with microservices. These microservices-based applications are deployed and overseen by the K8s container orchestrator, acting as the abstraction layer enabling the portability of CNFs and other containerized applications across diverse infrastructure. Service providers require an orchestration solution capable of managing both virtual network functions (VNFs)/CNFs (implemented as microservices) and oversee the host K8s clusters.

Service providers might need to accommodate scenarios where the session management function (SMF) and user plane function (UPF) simultaneously cater to multiple network slices, while also addressing cases where distinct SMFs and UPFs are allocated for each network slice. In Figure 9.6, the slices shown could be logical as well as physical. Moreover, the changing dynamics of network and slice performance may necessitate a transition from shared to dedicated instances to fulfill SLAs. For example, the core can have slices for eMBB, URLLC and MIoT – these slices could be on same CNFs (SMF/UPF) or could have dedicated CNFs.

Figure 9.6: Core slicing.

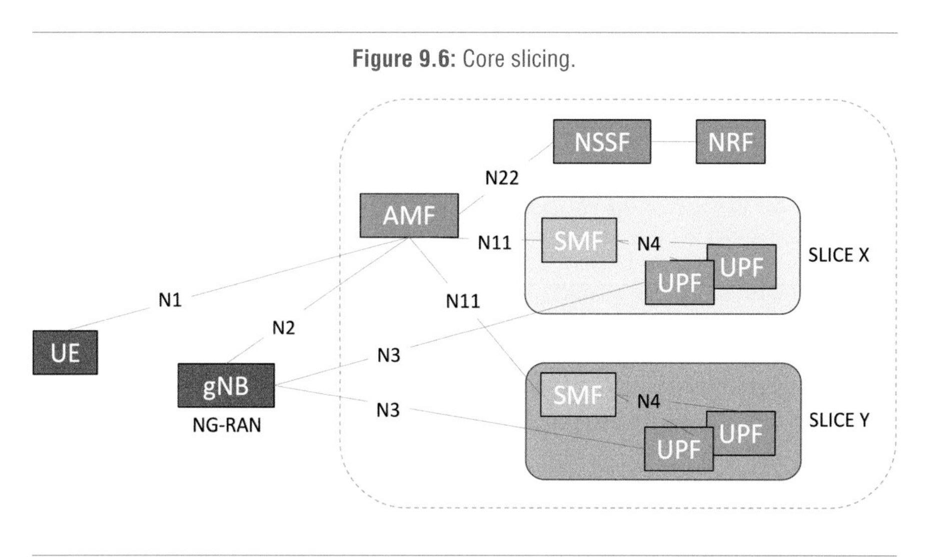

As defined by 3GPP, a UE can access up to eight slices. A PDU session can belong to only one slice. The same AMF serves as anchor for all the slices accessed by a given UE. In the 5G core, slice information is passed in multiple legs of signaling during the session establishment. A slice is identified by S-NSSAI (Single Network Slice Selection Assistance Information) and is made up of the following:

- **SST (slice/service type)**: This will define the expected behavior of the network slice in terms of specific features and services. Standardized SST values include eMBB (enhanced mobile broadband), URLLC (ultra reliable low latency communications) and MIoT (massive Internet of Things).
- **SD (slice differentiator)**: This is optional information that complements the slice/service type(s) and is used as an additional differentiator if multiple network slices carry the same SST value.

PDU Session information in SMF, AMF, and RAN has 1–1 association between PDU session and S-NSSAI. In the 5G core, NSSF function selects the network slice(s) to be used for a UE, based on the requested and subscribed S-NSSAI values.

9.4.3 Transport slicing

Segment routing enables service providers to support network slicing without any additional protocol, other than the SR IGP extensions. The network, in a distributed and entirely automated manner, can share a single infrastructure resource along multiple virtual service (slices).

For example, one slice can be optimized continuously for low-cost transport; a second slice can be optimized continuously for low-latency transport; a third slice can be orchestrated to support disjoint services, etc. The optimization objective of each of these slices is programmable by the operator.

The segment routing specification already contains the various building blocks required to create network slices. These include the following:

- SR policy with flexible algorithm
- TI-LFA with protection in the slice underlay
- SR VPN
- SR Service Programming using NFV
- QoS using DiffServ
- Stateless network slice identification

- Service provisioning/orchestration at the controller
- Operation, administration and management (OAM) performance management (PM).

Each of these foundational elements operates autonomously. Their capabilities can be combined to meet the service provider's need for network slicing. The orchestration of these building blocks into a slicing service is crucial, with a pivotal role played by an external controller.

The following sections elaborate on the attributes of some of these building blocks for network slicing.

9.4.3.1 Segment routing policy

A segment routing (SR) policy allows a headend node to steer a packet flow along any path without creating intermediate per-flow states. The headend node steers a flow into a segment routing policy (SR policy). This allows operators to enforce low-latency and/or disjoint paths, regardless of the normal forwarding paths. The SR policy can support various optimization objectives. The optimization objectives can be instantiated for the IGP metric or the TE metric or the latency extended TE metric.

In addition, an SR policy is able provide or define various constraints, including inclusion and/or exclusion of TE affinity, IP address, SRLG, admin-tag, maximum accumulated metric (IGP, TE, and latency), maximum number of SIDs in the solution SID-List, maximum number of weighted SID-lists in the solution set, diversity to another service instance (e.g., link, node, or SRLG), etc. depending on the vendor implementation.

The following subsection describes the SR flexible algorithm feature and how SR policy can utilize this feature.

9.4.3.2 Flex-algorithm

The flexible algorithm enhances the SR policy solution by introducing additional segments with properties distinct from those of the IGP prefix segments. Flex Algo introduces customizable, user-defined segments to the SRTE toolbox. More specifically, it facilitates the association of "intent" with prefix SIDs. The IGP-based flex-algorithm solution enables IGPs to compute paths constrained by the "intent" represented by the flex algorithm.

The flex-algorithm has the following attributes:

- Algorithm associated to the SID a specific TE intent expressed as an optimization objective (an algorithm).
- Flexibility includes the ability of network operators to define the intent of each algorithm they implement.
- By design, the mapping between the flex-algorithm and its meaning is flexible and is defined by the user.
- Flexibility also includes ability for operators to make the decision to exclude some specific links from the shortest path computation.

A network slice can be created by associating a Flexible-Algorithm value with the Slice via provisioning. Flex Algo leverages SR on-demand next hop (ODN) and Automated Steering for intent-based instantiation of traffic engineered paths described in the following sub-sections. Specifically, as specified in [RFC8402] the IGP Flex Algo Prefix SIDs can also be used as segments within SR Policies thereby leveraging the underlying IGP Flex Algo solution.

9.4.3.3 On-demand SR policy

Segment routing on-demand next-hop (ODN) functionality enables on-demand creation of SR policies for service traffic. Using a path computation element (PCE), end-to-end SR policy paths can be computed to provide end-to-end segment routing connectivity, even in multi-domain networks running with or without IGP flexible-algorithm.

The on-demand next-hop functionality provides optimized service paths to meet customer and application SLAs (such as latency, disjointness) without any pre-configured TE tunnel and with the automatic steering of the service traffic on the SR policy without a static route, autoroute-announce, or policy-based routing. With this functionality, a network service orchestrator can deploy the service based on their requirements. The service head-end router requests the PCE to compute the path for the service and then instantiates an SR policy with the computed path and steers the service traffic into that SR policy.

If the topology changes, the stateful PCE updates the SR policy path. This happens seamlessly, while TI-LFA protects the traffic in case the topology change happened due to a failure.

9.4.3.4 Automatic steering

Automatically steering traffic into a network slice is one of the fundamental requirements of slicing. This is made possible by the "automated steering" functionality of SR. Specifically, SR policy can be used to traffic engineer paths within a slice, "automatically steer" traffic to the right slice and connect IGP flex-algorithm domains sharing the same "intent".

A headend can steer a packet flow into a valid SR policy within a slice in various ways using the SR policy constraints. Incoming packets can have an active SID matching a local binding SID (BSID) at the headend.

- Per-destination steering: incoming packets match to BGP/service route which recurses on an SR policy with color attribute.
- Manual/automated per-flow steering: incoming packets match or recurse on a forwarding array of where some of the entries are SR policies.
- Policy-based steering: incoming packets match a routing policy which directs them on an SR policy.

9.4.3.5 Inter-domain considerations

The network slicing needs to be extended across multiple domains such that each domain can satisfy the intent consistently. SR has native inter-domain mechanisms, e.g., SR policies are designed to span multiple domains using a PCE based solution.

An edge router upon service configuration automatically requests to the segment routing PCE an inter-domain path to the remote service endpoint. The path can either be for simple best-effort inter-domain reachability or for reachability with an SLA contract and can be restricted to a network slice. The SR native mechanisms for inter-domain are easily extendable to include the case when different IGP flex-algorithm values are used to represent the same intent.

- Provider 1 (SP1) may use flex-algo 128 to indicate a low latency slice in domain1 service.
- Provider 2 (SP2) may use flex-algo 129 to indicate a low latency slice in domain2 service.

When an automation system at a PE1 in the SP1 network configures a service with the next hop (PE2) in the SP2 network, the SP1 contacts a path computation element (PCE) to find a route to the PE2. In the request, the PE1 also indicates the intent (i.e., the flex-algo 128) in the PCEP message.

As the PCE has a complete understanding of both domains, it can understand that the path computation in Domain1 needs to be performed for Algorithm 128 and the path computation in Domain2 needs to be performed for Algorithm 129 (i.e., in the low latency network slice in both domains).

9.4.4 Service slicing

Virtual (VPNs) provide a mean for creating a logically separated network to a different set of users access to a common network. Segment routing is equipped with the rich multi-service virtual private network (VPN) capabilities, including layer 3 VPN (L3VPN), virtual private wire service (VPWS), virtual private LAN service (VPLS), and Ethernet VPN (EVPN). The ability of segment routing to support different VPN technologies is one of the fundamental building blocks for creating slicing an SR network.

9.4.4.1 Stateless SR service chaining

An important part of the slicing is the orchestration of virtualized service containers. SR service chaining describes how to implement service segments and achieve stateless service programming in SR-MPLS and SRv6 networks.

It introduces the notion of service segments. The ability of encoding the service segments along with the topological segment enables service providers to forward packets along a specific network path, but also steer them through VNFs or physical service appliances available in the network.

In an SR network, each of the service, running either on a physical appliance or in a virtual environment, is associated with a segment identifier (SID) for the service.

These service SIDs are then leveraged as part of a SID-list to steer packets through the corresponding services. Service SIDs may be combined with topological SIDs to achieve service programming while steering the traffic through a specific topological path in the network.

In this fashion, SR provides a fully integrated solution for overlay, underlay and service programming building blocks needed to satisfy network slicing requirements.

9.5 OAM Aspects of Network Slice

There are various OAM aspects that are critical to satisfy slicing requirements. These include but are not limited to the following:

- Measuring per-link TE matric.
- Flooding per-link TE matric.
- Taking TE matric into account during path calculation.
- Taking TE matric bound into account during path calculation.
- SLA monitoring.

The service provider can monitor each SR policy in a slice to monitor SLA offered by the policy using techniques like a performance measurement toolset. This includes monitoring end-to-end delays on all ECMP paths of the policy as well as monitoring traffic loss on a policy. Remedial mechanisms can be used to ensure that the SR policy conforms to the SLA contract.

9.6 QOS

Segment routing relies on MPLS and IP differentiated services. Differentiated services enhancements are intended to enable scalable service discrimination in the Internet without the need for per-flow state and signaling at every hop.

This architecture for implementing scalable service differentiation in the network is composed of many functional elements implemented in network nodes, including a small set of per-hop forwarding behaviors, packet classification functions, and traffic conditioning functions including metering, marking, shaping, and policing.

The DiffServ architecture achieves scalability by implementing complex classification and conditioning functions only at network boundary nodes, and by applying per-hop behaviors to aggregates of traffic depending on the traffic marker. Specifically, the node at the ingress of the DiffServ domain conditions, classifies, and marks the traffic into a limited number of traffic classes. The function is used to ensure that the slice's traffic conforms to the contract associated with the slice.

Per-hop behaviors are defined to permit a reasonably granular means of allocating buffer and bandwidth resources at each node among competing traffic streams. Specifically, per class scheduling and queuing control mechanisms are applied at each IP hop to the traffic classes depending on packet's marking. Techniques such as queue management and a variety of scheduling mechanisms are used to get the required packet behavior to meet the slice's SLA.

9.7 Stateless Network Slice Identification

Some use-cases require a slice identifier (SLID) in the packet to provide differentiated treatment of the packets belonging to different network slices. The network slice instantiation using the SLID in the packet is required to work with the building blocks described in the previous sections. For example, the QoS/ DiffServ needs to be observed on a per slice basis. The slice identification needs to be topologically independent and stateless.

Stateless slice identification in SRv6 describes a stateless encoding of slice identification in the outer IPv6 the header of an SRv6 domain. When an ingress PE receives a packet that traverses the SR domain, it encapsulates the packet in an outer IPv6 header and an optional SRH. Based on a local policy of the SR domain, the flow label field of the outer IPv6 header carries the SLID. Specifically, the SLID is added in the eight most significant bits of the flow label field of the outer IPv6 header. The remaining 12 bits of the flow label field are set as described in section 5.5 of [RFC8754] for inter-domain packets.

Based on the local policy of the SR domain, the RFC also uses one of the bits in the traffic class field of the outer IPv6 header to indicate that the entropy label contains the SLID. Stateless slice identification in SR-MPLS describes a similar stateless encoding of slice identification in the SR-MPLS domain. Specifically, the RFC extends the use of the entropy label to carry the SLID. The number of bits to be used for encoding the SLID in the entropy label is governed by a local policy of the SR domain. Based on the local policy of the SR domain, the draft uses one of the bits in the TTL field of the entropy label to indicate that the entropy label contains the SID.

The network slicing mechanism described using the entropy label works seamlessly with the building blocks described in the previous sections. For example, the slice identification is independent of topology and the network's QoS/DiffServ policy. It enables scalable network slicing for SR-MPLS overlays.

9.8 Orchestration at the Controller

A controller plays a vital role in orchestrating the SR building blocks discussed in previous sections to create network slices. The controller also performs admission control and traffic placement for slice management at the transport layer.

The SDN friendliness of the SR technology becomes handy to realize the orchestration. The controller may use PCEP or Netconf to interact with the routers. Most routers implement Yang model for SR-based network slicing.

9.9 End-to-end Slice

End-to-end slice support is imperative to cater to diverse B2B enterprise use cases and meet varying network performance requirements in 5G. Achieving true end-to-end network slicing necessitates a programmable 5G network platform.

The automation of cross-domain slice management and operations plays a pivotal role in realizing economies of scale. Cross-domain network slice orchestration becomes pivotal for enabling end-to-end slicing. The orchestration of slicing within the cloud-native 5G core is facilitated by the 5G core domain orchestrator. Seamless interaction with the Kubernetes orchestrator is imperative for the 5G core domain orchestrator (Figure 9.7).

Figure 9.7: End to end slicing orchestration.

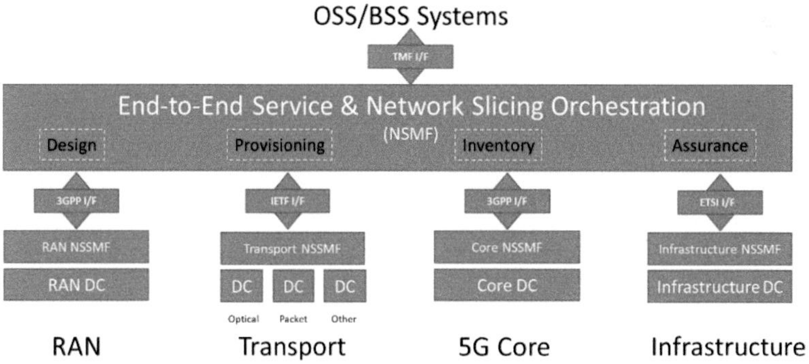

9.10 Service Level Objectives

Service level objectives (SLOs) can be categorized as "directly measurable objectives" or "indirectly measurable objectives" when provisioning the network slice. Objectives such as guaranteed minimum bandwidth, guaranteed maximum latency, maximum permissible delay variation, maximum permissible packet loss rate, and availability are "directly measurable objectives", while "indirectly measurable objectives" include security, geographical restrictions, maximum occupancy level objectives, etc. The SLO objectives are defined by customer or service type which is deployed in the customer network.

9.11 Cross-domain Network Slice Orchestration

The high-level network slice management framework published by 3GPP outlines four key management functions for network slicing:

- Communications service management function (CSMF)
- Network slice management function (NSMF)
- Network slice subnet management function (NSSMF)
- Network function management function (NFMF).

The CSMF is a higher-layer OSS/BSS capability aimed at performing customer order management, and the NFMF is for the application-level management of VNFs, PNFs and CNFs.

The NSMF will perform cross-domain network slice orchestration using the domain-level slice management functions, with each network domain having its own NSSMF.

This architecture allows for the instantiation and configuration of network slice resources for each of the use case types (eMBB, mMTC or uRLLC) in each subnet or domain, dictated by the end-to-end network slice intent and governed by the end-to-end slice orchestrator (Figure 9.8).

It is expected that additional new innovative use cases will be identified to fit the needs of enterprises that require specific levels of service. In some cases, the combination of slice parameters will be new and unique, while in other cases, multiple services will share common characteristics, albeit with differences in the target infrastructure.

It is anticipated that the identification of novel use cases tailored to meet the specific service requirements of enterprises will happen. Some instances may introduce innovative combinations of slice parameters, while others might

Figure 9.8: Cross domain slicing orchestrators.

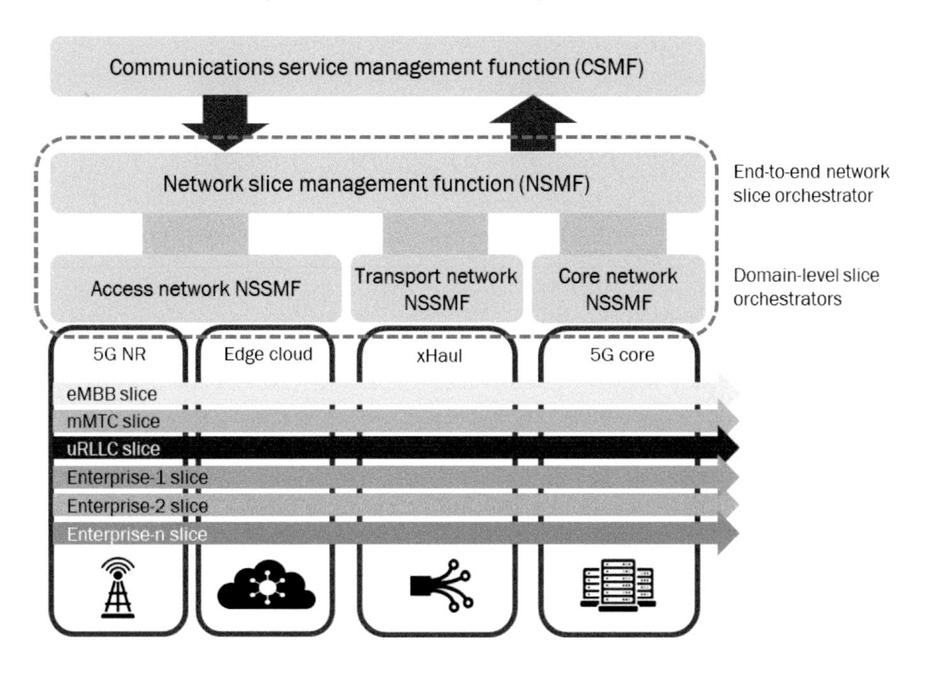

involve multiple services with shared attributes, albeit with distinctions in the targeted infrastructure.

Consequently, it is envisaged that the network slice management function (NSMF) will uphold a repository of slice templates for prevalent use cases. This repository will facilitate the sharing and swift provisioning of slices owing to their similarities.

The NSSMFs are likely to be built as extensions of the existing domain orchestrators (e.g. 5G core orchestrator) and the domain controllers (such as the SDN controller for the transport network).

The cross-domain nature of the NSMF has been discussed at a high level in various standards bodies (most notably by ETSI, where it is referred to as the Service Orchestrator). However, the concept of slice-based service orchestration and lifecycle management is still new, and the standards are still emerging.

9.11.1 Design principle of end-to-end network slice orchestration solution

In addition to having an end-to-end cross-domain capability, the network slice orchestration solution must be developed based on the following design principles:

9.11.1.1 Multi-layer architecture

A fundamental aspect of this solution is to deliver multi-layer control both within specific domains and across various domains. Through a multi-layer orchestration approach, abstraction and operational boundaries are ensured in a highly distributed mobile network architecture. This architecture encompasses multiple domains such as networking, data centers, and clouds, including RAN, WAN, edge, and core components.

The lower-layer domain orchestrators are responsible for individual domain-level orchestration and software-defined control within each specific domain. Simultaneously, the cross-domain orchestrator functions as the top level orchestrator, offering higher-layer abstraction and facilitating end-to-end cross-domain service orchestration.

9.11.1.2 Model or intent driven

The mobile network environment is characterized by a diverse array of network infrastructure, encompassing traditional physical network functions, VM-based network functions, and the latest container network functions designed for 5G standalone (SA). Each component within this spectrum exhibits varying levels of maturity. Consequently, the orchestration solution needs to endorse a model-based approach, offering a high degree of customization to accommodate diverse permutations of network functions and meet specific requirements. This approach empowers service providers to innovate and develop new services and use cases without being overly dependent on the individual components of the underlying network.

9.11.1.3 Platform-oriented and open API-driven

Opting for a platform-centric approach facilitates the construction of the solution in a modular fashion, leveraging a microservices-based architecture.

This design choice ensures ease in introducing new capabilities, enhances the composability of the solution, and enables service improvements with minimal disruption to the run-time environment.

The solution's adherence to open, industry-standard APIs is crucial for seamless solution composition and seamless integration with third-party software components, fostering best-of-breed orchestration. Furthermore, the platform should be developed using a carefully curated set of tools for effectively managing DevOps and continuous integration/continuous deployment (CI/CD) pipelines.

9.11.1.4 Multi-vendor support

The cross-domain orchestration solution for network slicing (NSMF) needs to possess multi-vendor capabilities, ensuring seamless integration with any third-party domain-level slice orchestrators (NSSMF). These domain-level orchestrators should maintain vendor-agnostic characteristics, capable of orchestrating networks comprising a variety of multi-vendor network equipment, virtual network functions (VNFs), and containerized network functions (CNFs).

As network virtualization advances, it is anticipated that certain network functions will be hosted in a hybrid-cloud environment, encompassing both private and/or public clouds. Consequently, there arises a necessity for multi-cloud slice orchestration capabilities, extending across the private cloud environment within service providers' data centers and public clouds.

9.11.2 Standards for network slice management

Various standards development organizations (SDOs) are actively engaged in initiatives to establish standards and specifications for the management and operations of network slices (Figure 9.9). The GSMA and the NGMN Alliance function as prime integrators, ensuring alignment and convergence of technology standards related to network slicing within the industry.

Within this framework, 3GPP is taking the lead in creating technical specifications for slicing in both the RAN and core domains. The IETF is addressing enhancements to the IP router protocol, including technologies such as segment routing and L3VPN, and collaboratively working with 3GPP to define interfaces between RAN and core networks for slice management.

Figure 9.9: Standardization environment of network slicing.

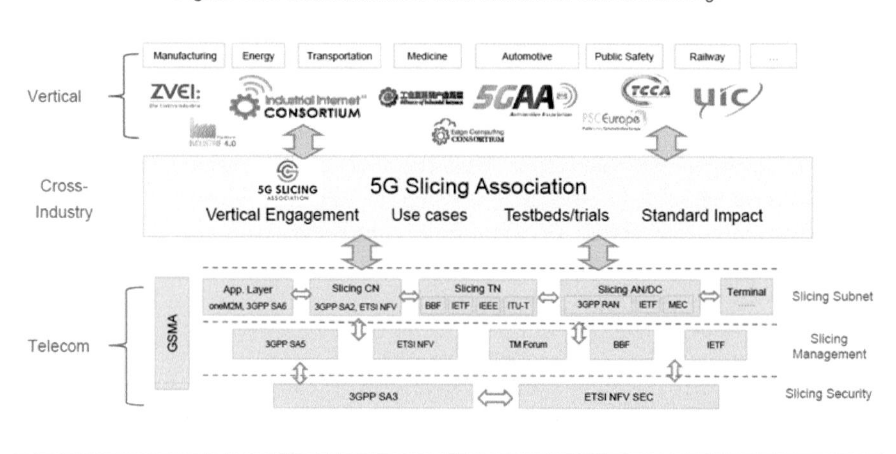

Responsibilities for technical specifications related to the management and orchestration of NFV-based and cloud-native 5G core rest with the ETSI ISG NFV. Simultaneously, the ETSI ISG ZSM is dedicated to cross-domain network slice management, determining how the end-to-end orchestrator interfaces with individual domain-level slice orchestrators. The O-RAN Alliance is actively developing specifications for Open RAN, with ongoing efforts to incorporate RAN slicing requirements.

The extensive involvement of numerous SDOs and organizations in developing comprehensive specifications for network slicing lifecycle management underlines the complex nature of the industry's undertaking. Deep collaboration among SDOs, service providers, vendors, and industry consortia will be imperative to achieve consensus and standardization.

CHAPTER

10

5G Security

Security represents a critical pillar of the 5th generation of wireless technology. Unlike its predecessors, 5G is not just about faster speeds; it introduces a comprehensive security framework to safeguard networks, devices, and data in an era of increasing connectivity. With its ultra-fast speeds and low-latency communication, 5G networks are poised to underpin the Internet of Things (IoT), autonomous vehicles, and critical infrastructure, making security paramount.

5G security encompasses various aspects, including robust encryption to protect data in transit, stringent authentication mechanisms, and secure device management. It also extends to safeguarding network functions, virtualized components, and the rapidly growing ecosystem of IoT devices. The implementation of AI and machine learning for threat detection and mitigation adds another layer of sophistication to 5G security.

In an age where cyber threats are constantly evolving, 5G security seeks to stay ahead by adopting a proactive stance, emphasizing resilience, and adhering to regulatory compliance. As the deployment of 5G networks continues to expand, the collaboration of network operators, service providers, manufacturers, and policymakers will be crucial in addressing the complexities of 5G security and ensuring that the benefits of this technology are realized securely and responsibly.

The network should be planned and deployed with a robust security framework in mind. Achieving this requires adopting a comprehensive approach that covers the protection of network components, functionalities, services, management, control systems, and data transmission. This approach combines elements of both a "risk-centric" and a "trust-centric" security model to ensure comprehensive security (Figure 10.1).

Figure 10.1: 5G security model.

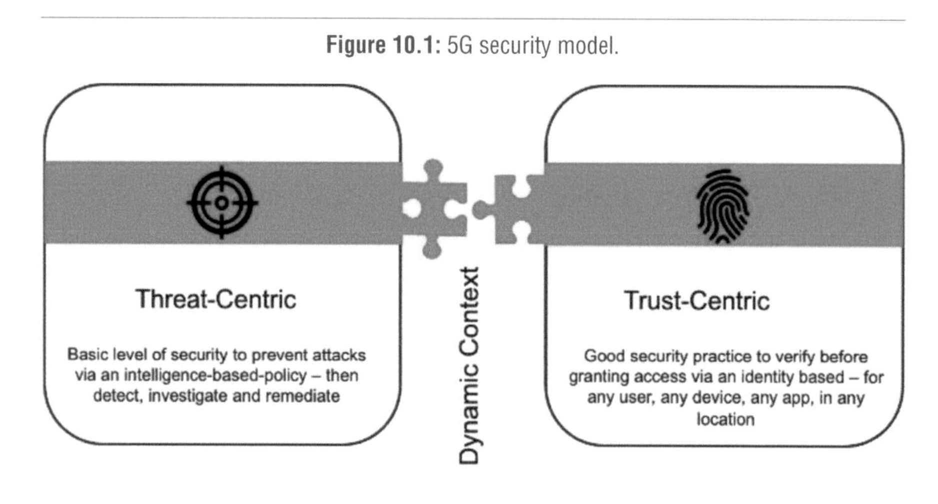

The threat-centric perspective is based on information provided by various 5G standardization groups and organizations like ETSI, 3GPP, 5GPPP, and ENISA. The 5G threat classification and asset exposure serve as inputs for constructing a 5G threat model.

On the other hand, our trust-centric approach prioritizes the protection of resources, including assets, services, workflows, and confidential information. In this approach, the network's physical location is no longer the primary factor in securing these resources. Instead, it involves the authentication and authorization of discrete functions before establishing a secure session with a resource, utilizing secure protocols.

5G necessitates the use of cloud-native solutions and CI/CD practices, where security measures are integrated into the CI/CD pipelines. This integration automates tasks like segmentation, monitoring, behavioral analysis, threat identification, forensic analysis, and containment.

Security is an integral part of the network design and deployment process, with security maturity evolving alongside the network and business development. Security configurations and policies are stored in source control, and any changes are managed within code branches. Build servers are responsible for deploying and testing proposed configuration modifications.

10.1 RAN and xHaul Security

In general, a distributed 5G RAN will have NFs at the edge of the network, potentially at less secure locations or sites with minimal physical security. RMI's network should ensure that these distributed NFs cannot access cryptographic keys protecting subscriber traffic, thereby protecting the traffic from an attacker physically compromising the site. Separation of RAN and core is critical to the evolution of 5G networks, and thus may pose hurdles in securing multiple use cases that have been important drivers for 5G development.

RAN interfaces introduce multiple security concerns. These include threats to privacy, availability, integrity, and confidentiality. Some of these threats are:

- User identity catching – aka IMSI catching
- Tracking user locations
- DDoS and Botnet attack on management plane
- Eavesdropping of the user and signaling planes
- DDoS to the core network using the signaling plane
- User traffic modifications.

To counter some of these threats, 3GPP ETSI has added the following security enhancements for 5G:

Air interface security: Initial messages are now encrypted, unlike 4G and air interface security; hence, more secure.

Subscriber security: Security in the 3GPP 5G standard enhances protection of subscriber privacy against false base stations, popularly known as IMSI catchers or Stingrays. This is achieved by enhancing security for UE and subscriber by using the SUPI to SUCI mechanism.

Fronthaul security: PDCP provides initial level security for the access stratum (AS) and non-access stratum (NAS).

New authentication framework: Network and device mutual authentication in 5G is based on primary authentication. The authentication mechanism has in-built home control, allowing the home operator to know whether the device is authenticated in a given network and to take the final call of authentication. In 5G Phase 1, there are two mandatory authentication options: 5G authentication and key agreement (5G-AKA) and extensible authentication protocol (EAP)-AKA' (that is, EAP-AKA'). Secondary authentication in 5G is meant for authentication with data networks outside the mobile operator domain. For this purpose, different EAP-based authentication methods and associated credentials can be used.

Figure 10.2 shows the 5G initial security enhancements.

Figure 10.2: 5G initial security enhancements.

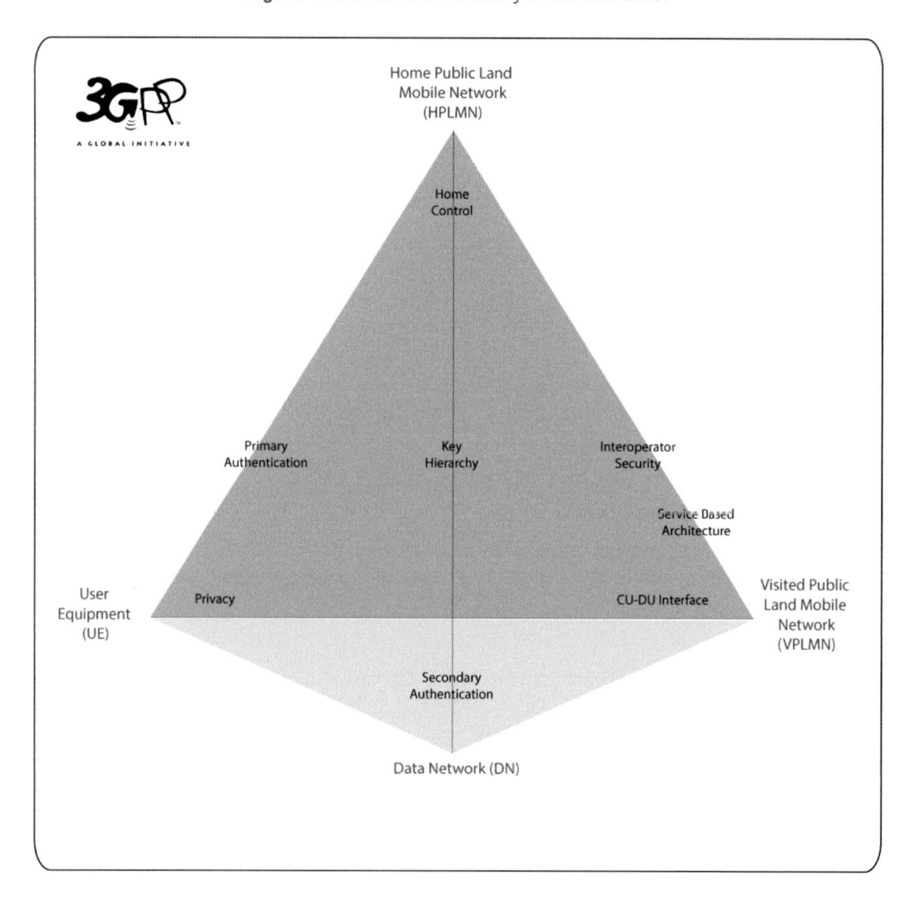

10.1.1 Options to secure RAN and xHAUL

xHaul will provide transport between RAN/fronthaul and CDC data centers using a next-generation IPv6 flat routing architecture. Flexible, scalable, simplified xHaul solutions result in delivery of increased mobile capacity and coverage, with the right QoS.

Confidentiality, integrity, ciphering, and replay protection can be achieved either by using inherent security controls or by introducing components that can help achieve these functionalities. The following are some security options.

10.1.1.1 Security matrix for non-SBI reference points

Table 10.1 depicts various reference points and requirements for integrity and ciphering according to 3GPP standards. The table depicts the integrity and ciphering planned for deployment and control options. As defined in the earlier trust model, a security control should be in place where traffic crosses a trust boundary. This table can help to select an appropriate control option and determine its placement in their network.

Table 10.1: Security matrix – non-SBI reference points.

UP/CP	Ref. Point	Interface	Protocol	Transport	3GPP standards		Optionality	Planned in Rakuten		3GPP control options		Comment
					Integrity	Ciphering	Full/Partial	Integrity	Ciphering	NDS/IP	Transport	
Control	N1	UE <> SMF	NAS-SM	SCTP/NGAP	M	O	Partial	☑	?	N/A	N/A	Inherint NAS security
Control	N1	UE <> AMF	NAS-MM	SCTP/NGAP	M	O	Partial	☑	?	N/A	N/A	Inherint NAS security
Control	N1	UE <> AMF	NAS-MM	SCTP/NGAP	M	O	Partial	☑	?	N/A	N/A	Inherint NAS security
Control	N2	CU-CP <> AMF	NGAP	SCTP	M	M	Full	☒	☒	IPSec	DTLS	
Control	F1-C	UE <> CU-CP	RRC	PDCP	M	O	Full	☑	☑	IPSec		Altiostar DU/CU
Control	F1-C	DU <> CU-CP	F1AP	SCTP	M	M	Full	☑	☑	IPSec	DTLS	Altiostar DU/CU
Control	E1	CU-CP <> CU-UP	E1AP	SCTP	M	M	Full	☒	☒	IPSec	DTLS	
User	F1-U	UE <> CU	SDAP	PDCP	M	M	Full	☒	☒	IPSec		
Control	Xn-C	CU-CP <> CU-CP	XnAP	SCTP	M	M	Full	☒	☒	IPSec	DTLS	
User	Xn-U	CU-UP <> CU-UP	GTP-U	UDP	M	M	Full	☒	☒	IPSec		
User	N3	CU-UP <> UPF	GTP-U	UDP	O	O	Full	☒	☒	IPSec		
User	N9	UPF <> UPF	GTP-U	UDP	O	O	Full	☒	☒	IPSec		
Control	N4-C	UPF <> SMF	PFCP	UDP	O	O	Full	☒	☒	IPSec		
User	N4-U	UPF <> SMF	GTP-U	UDP	O	O	Full	☒	☒	IPSec		

Note: *Operator optionality means 3GPP states that the use of cryptographic solutions to provide integrity, confidentiality and anti-replay protection is an operator's decision.*

- O: Vendor support optional
- M: Vendor support mandatory.

10.2 IP Transport Security

10.2.1 IPSec via SecGW

One option to secure interfaces such as Xn-C *(based on the limitation mentioned)*, N2, N3, and N4, are to deploy SecGWs at relevant locations.

10.2.2 MACSec

MACSec provides point-to-point line encryption. It uses layer 2 IEEE 802.1AE standard for encrypting packets between two MACsec-capable routers. MACsec secures data on physical media, and is not specific to a particular type of communication or reference point. MACSec also helps prevent layer 2 security breaches, including packet sniffing, packet eavesdropping, DOS attack, tampering, MAC address spoofing, ARP spoofing, and so on.

10.2.3 TACACS integration

All access to console, remote https/ssh must be authenticated by an AAA system. The AAA device should support administration using TACACS+ and RADIUS security protocols to control and audit network devices. The network devices are configured to query the AAA for authentication and authorization of device administrator actions and send accounting messages for the AAA device to log the actions. It facilitates granular control of who can access which network device and change the associated network settings.

10.2.4 Syslog integration

Event logging provides visibility into the operation of a 5G SA device, and the network into which it is deployed. By default, each log message generated by the transport network device may be assigned one of eight severities that range from level 0 (Emergencies) through level 7 (Debug). Unless required, avoid logging at level 7. Logging at level 7 produces an elevated CPU load on the device that can lead to device and network instability.

To provide an increased level of consistency when collecting and reviewing log messages, transport devices need to statically configure a logging source interface, if applicable. This ensures that the same IP address appears in all logging messages sent from an individual device.

Configuration of logging timestamps helps correlate events across network devices. It is critical to implement a correct, consistent logging timestamp configuration to ensure that we are able to correlate logging data. Logging timestamps should be configured to include the date and time with millisecond precision, and to include the time zone in use on the device.

These syslog messages also are used to review and identify malicious traffic.

10.3 Telco Cloud and MEC Security

5G heralds the next phase of mobile evolution, offering improved connectivity, reliability, and minimal latency. However, this progress introduces a range of new risks and potential attack vectors that service providers must take into account when strategizing their deployments. While security in 5G is comprehensive, this book focuses on specific aspects of security.

- Virtual infrastructure security
- MEC security
- Network slicing security
- Security management
- 3GPP trust model for core.

10.3.1 Virtualization threats

The security risks related to virtual network functions (VNFs) arise from the convergence of security threats in both physical networking and virtualization technologies.

10.3.1.1 Kernel escape (Figure 10.3)

The security risks linked with VNFs result from the amalgamation of threats related to virtualization technologies. With type 1 hypervisors, guest OS instances operate within controlled, sandbox-like environments. An escape attack involves breaching this separation, allowing an unauthorized user to gain access to the hypervisor. This can serve as a launching point for lateral movement within the network or compromising other co-resident guest images on the same device. As operators embrace container strategies, these attacks become more feasible due to the diminishing natural isolation of virtual machines.

Figure 10.3: Kernel escape.

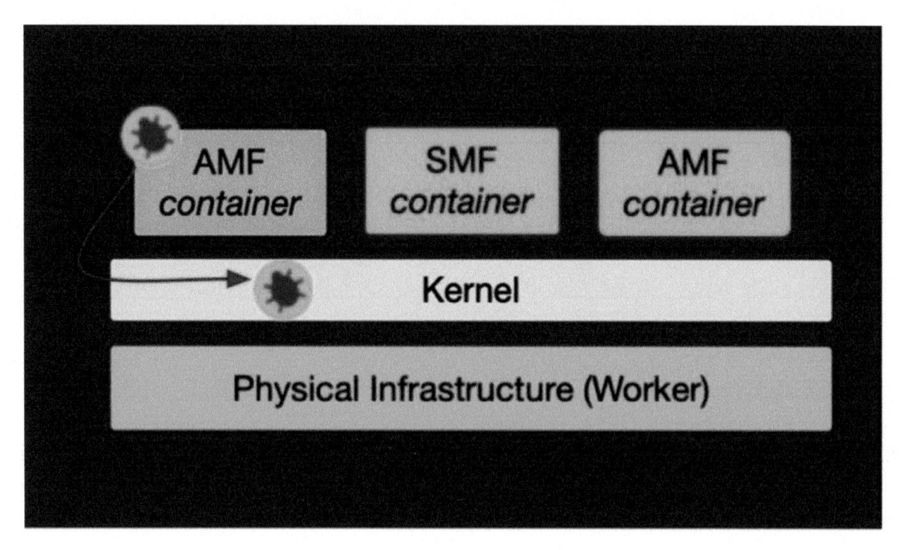

Countermeasures:

- Vigilantly monitor hypervisor vulnerabilities and establish a robust vulnerability management system that filters out extraneous information, offering comprehensive insights into the vulnerability status of your assets.
- For containers, make sure to deploy an admission controller. This controller should possess the capacity to enforce security contexts and policies, including the deactivation of RunAsRoot and the prohibition of privileged containers.
- Validate the compliance of your NFVi (network functions virtualization infrastructure) platform with recognized standards such as CIS or NIST. Conduct routine automated scans to gain visibility into your environment's compliance status.

10.3.1.2 Lateral movement (Figure 10.4)

Lateral movement entails cyber attackers employing techniques to navigate a network in search of more valuable resources. In a virtualized environment, connectivity is considerably simpler compared to physical cable setups. This reduced complexity can lower the incentive to implement security controls, potentially leaving the network vulnerable to attacks.

Figure 10.4: Lateral movement.

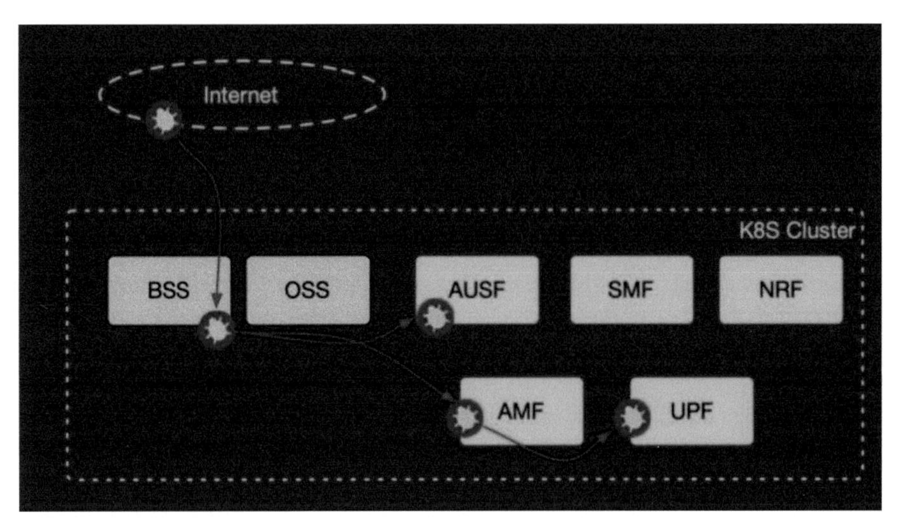

Countermeasures:

- Establish a comprehensive network security policy for virtual machines, such as utilizing network security groups in Openstack.
- For containers, enforce the use of an admission controller to ensure that containers cannot be launched without a default deny network policy.
- Maintain a robust endpoint security solution supported by a trusted threat research team.

10.3.1.3 VM/container sprawl

Virtual machines and containers can be effortlessly generated, duplicated, and transferred to physical servers. Malicious actors can exploit this technique to establish persistence and propagate within the network while evading detection. This can introduce complexity in terms of auditing and security monitoring, potentially leading to a loss of network control.

Countermeasures:

- Enhance your OSS (operations support system) with features that enforce a stringent VM/container build process and change management.

- Implement a middleware portal within your OSS to oversee your virtualized assets. This portal can be employed to establish policies for monitoring, managing, or removing unused and unclaimed assets.
- Employ a privileged access management solution to secure and audit management sessions for your hypervisor.

10.3.1.4 Data breaches

Virtualization and containers have the capability to store data with varying levels of sensitivity classification. The trust levels of virtual machines are determined by the classification of the data they contain. Often, virtual machines housing sensitive data like passwords, personal information, bash profiles, and encryption keys are placed in the same trust category as other virtual assets. This poses a significant risk because this data can be much more easily transferred than the physical hard drive of a server. The risk is further amplified when virtual machine snapshots are used, as they capture the contents of memory at the time of the snapshot. Additionally, during VM migration, residual data may be exposed in their previous locations.

Countermeasures:

- Establish a data classification process specifically tailored to your virtualized assets, with special attention to business support systems (BSS).
- Deploy a network as a sensor solution to identify irregularities in traffic patterns.
- Implement data encryption at rest to protect sensitive information.
- Utilize a vault solution for managing organizational secrets, allowing developers to securely store passwords, keys, bash profiles, and more.
- Ensure that all assets are configured to transmit logs to a security operations center (SOC).
- Implement AAA mechanisms for all assets.
- Promote the use of golden images wherever feasible, which come with integrated support for OSS, SOC, and AAA servers.

10.3.1.5 Resource exhaustion

In a virtualized setting, software that heavily utilizes specific physical server resources can deplete those resources, potentially impacting the availability of virtual machines (VMs). This situation arises due to resource contention within the shared environment of a physical server, especially when multiple VMs are simultaneously running the same resource-intensive software, such as anti-virus scanning.

Countermeasures:

- Utilize resource pools to restrict the utilization of resources by your virtualized assets, based on data classification.
- Enforce policies that mandate the definition of resource limits for virtual assets. In Kubernetes, this control can be implemented through security contexts or an admission controller.

10.3.2 MEC threats

The MEC system should ensure a secure environment for executing services involving various stakeholders, including users, network operators, third-party application providers, application developers, content providers, and platform vendors. The following sections outline threats and specific measures to address them within the MEC context.

10.3.2.1 False or rogue MEC gateway

The open nature of edge gateways, where even user-owned devices can transform into fully functional participants (e.g., personal cloudlets, TV smart-boxes, etc.), creates a scenario where malicious actors might deploy their own gateway devices. This threat yields similar consequences to a man-in-the-middle attack.

Countermeasures:

- The MEC platform should authenticate all MEC application instances and only grant them access to authorized information. MEC specifications require the use of OAuth 2.0 for authorization of access to RESTful MEC service APIs defined by ETSI ISG MEC. The implementation of the OAuth 2.0 authorization protocol should adhere to IETF RFC 6749 [31] using the client credentials grant type and employ bearer tokens as per IETF RFC 6750 [32].
- Ensure the encryption of data in transit, particularly for data containing sensitive information.
- Maintain a robust vulnerability management solution for edge applications and the hypervisor to monitor vulnerabilities effectively, filtering out extraneous data and providing pertinent vulnerability information.
- Implement a trust platform manager to allow only trusted hardware within your network.
- Enforce transport security for EDGE-1-9 interfaces.

10.3.2.2 Edge node overload

This threat pertains to attacks on edge networks that disrupt the immediate vicinity of the affected networks, whether at a local level or targeting specific services. Overload can occur through the inundation of the edge node with requests or traffic directed at this component, often initiated by a particular mobile app or IoT device.

Countermeasures:

- Deploy DDoS protection at the service exposure layer.
- Establish segmentation policies between MEC applications to minimize the attack surface.
- Secure the physical access to locations housing MEC infrastructure.
- Implement TLS on each interface to safeguard data confidentiality and integrity.
- The MEC platform should authenticate all MEC application instances and grant them access only to authorized information using OAuth.
- Implement real-time security monitoring to detect anomalies in traffic patterns.

10.3.2.3 Abuse of edge open application programming interfaces (APIs)

The exploitation of vulnerabilities in MEC (multi-edge computing) applications leads to the misuse of open APIs. The necessity for open APIs in MEC primarily revolves around supporting federated services and enabling interactions with various providers and content creators. This threat encompasses potential issues such as DoS (denial of service), man-in-the-middle attacks, malicious mode problems, privacy breaches, and manipulation of virtual machines (VMs).

Countermeasures:

- Ensure that API interfaces adhere to the Common API Framework (CAPIF) for compliance.
- In a containerized environment, employ a service mesh to implement TLS and OAuth on service-based interfaces (SBIs).
- Utilize an API gateway that offers API security or incorporate API security as a sidecar for protecting APIs.

10.4 5G Core Security

3GPP defines a trust model for the 5G core that provides operators with guidance on how to implement security on their core. Trust within the network is considered to decrease the further one moves from the core.

Figure 10.5: 3GPP trust model.

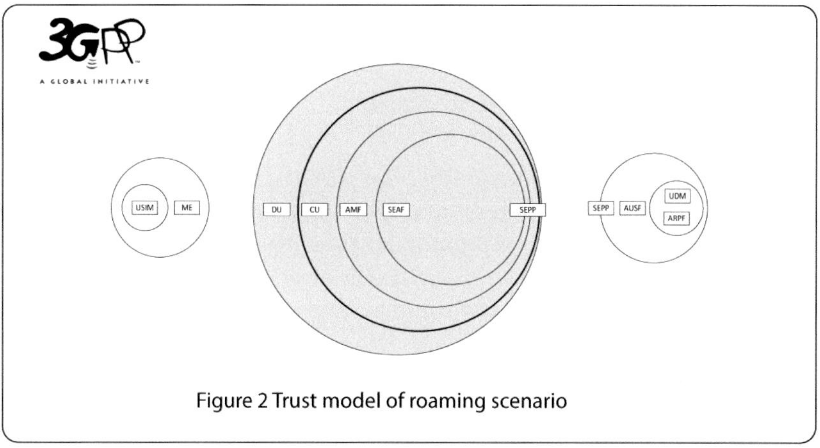

Figure 2 Trust model of roaming scenario

Reference: Anand R. Prasad, Sivabalan Arumugam, Sheeba B and Alf Zugenmaier, "3GPP 5G Security", Journal of ICT Standardization (River Publishers, Vol. 6, Iss. 1&2).

It is left up to the operator to decide how to define each of the rings displayed in the Figure 10.5. In most cases, the following controls are utilized to deploy this trust model at the core:

- Network segmentation
- Micro segmentation
- Smart contracts
- Kernel isolation
- Network firewalls
- Web application firewalls
- API security
- Mutual authentication with OAuth
- Transport layer security (TLS).

10.5 Network Slicing Security

10.5.1 Network slicing threats

Slices represent virtualized, self-contained logical networks responsible for facilitating network communication between user equipment and 5G services.

These slices are end-to-end communication channels that are virtually combined and associated with resources within the virtualized physical network infrastructure. The subsequent threats are associated with network slicing

10.5.2 Exploitation of poorly configured systems

Often categorized as a vulnerability, the threat arises from the exploitation of misconfigured or inadequately configured systems. The exploitation of such misconfigurations, typically unintentional, provides an opening for threat actors to access critical network assets or launch attacks. Configuration errors can manifest at various stages of the solution implementation lifecycle, including product installation and maintenance. Examples encompass poorly configured APIs, network functions, access control rules, network slices, administrative privileges, virtualized environments, traffic isolation, edge nodes, orchestration software, firewalls, and more.

Countermeasures:

- Implement a configuration management system for network infrastructure, firewalls, orchestration software, and edge nodes.
- Conduct regular automated configuration hardening checks to ensure systems comply with security configurations.

10.5.3 Exploitation of poorly designed architecture and planning (network, services, and security)

Termed as unintentional damage, this threat arises from insufficient design, planning, or incorrect adaptation, stemming from the myriad options and features inherent in the technology, from its original conception to actual implementation. The complexity involved and the challenges in achieving an optimal architecture, robust security measures, and effective operating procedures can result in subpar design and implementation. Design weaknesses present opportunities for malicious actors to exploit. By identifying inadequately implemented or protected features, malicious actors can exploit vulnerabilities and inject malware into the core network.

Countermeasures:

- Establish a process for architectural reviews conducted by an advisory board consisting of all relevant stakeholders, including those from cloud, infrastructure, security, and other pertinent areas.
- Implement a process to verify that implementations align with the intended designs.

10.6 Security management

As MEC deployments and partnerships with cloud providers become more prevalent in the context of 5G deployments, the operational landscape is becoming increasingly complex. It is imperative to integrate security seamlessly into automation systems to safeguard against a growing array of sophisticated cyber threats. Checklist outlining key steps operators should take to prepare for and defend against the expected surge in advanced attacks, given the expanded threat surface.

- Element management system (EMS) layer, for situational awareness
- Anomaly detection
- Trusted integrity measurements
- Certificate management
- Log auditing
- Network element (NE) vulnerability management
- End-to-end (e2e) security operation center (SOC)
- AI-based threat analysis and detection
- Security orchestration and network element (NE) vulnerability management.

CHAPTER

11

Migration from 4G to 5G

11.1 5G Deployment Options

3GPP has defined 5G to be cloud native right from its inception. There is no real dependency on previous generations for a 5G deployment. However, in reality, operators would want a gradual transition to 5G. The move to 5G has to encompass an end to end migration. The IP transport evolution based on SR-MPLS, as detailed in Chapter 5, plays an important role in assuring that latency and bandwidth requirements will be met on migration to 5G. Operators will need to implement software-defined networking (SDN) to enable network flexibility and scalability, supporting dynamic service provisioning and slicing. Automation and orchestration capabilities need to be factored earlier in the migration to assure that the 5G core elements can dynamically scale up and down as per the traffic requirements. Still the most important aspect remains to be the migration of packet core. In this chapter we cover the migration aspects of the packet core.

3GPP has gone a step forward to define two flavors of 5G – SA (standalone) and NSA (non-standalone). Technical details of both these options have been covered under Chapter 7. In this chapter we look at the options available under SA and NSA that an operator should consider while deploying 5G (Figure 11.1).

There are three variations of SA:

- Option 1 using EPC and LTE eNB access (pure 4G)
- Option 2 using 5GC and NR gNB access

- Option 5 using 5GC and LTE ng-eNB access.

Similarly, there are three variations of NSA -

- Option 3 using EPC and an LTE eNB acting as master and NR en-gNB acting as secondary
- Option 4 using 5GC and an NR gNB acting as master and LTE ng-eNB acting as secondary
- Option 7 using 5GC and an LTE ng-eNB acting as master and an NR gNB acting as secondary.

Most operators will go for NSA to start their 5G journey as they can still leverage their existing 4G network. They can combine the existing LTE radio access with the newly introduced NR access and connect to the existing 4G core itself. Although the user experience will depend on the radio technology used, this allows for a first taste of 5G. Note that the core might need to be migrated to CUPS from traditional SGW/PGW to achieve the expected throughput rates.

11.1.1 Greenfield approach

A greenfield operator can adopt a transformative deployment strategy by assessing the preparedness of data centers, IP backbone, and automation capabilities as part of their initial 4G rollout plan. This strategy involves virtualizing all network components, including radio network elements, adopting automated deployment and distributing them across various data centers during the 4G deployment phase. This approach makes the transition to 5G smoother and more efficient, as they already have the necessary infrastructure to support slicing, edge computing use cases, and other 5G applications. Additionally, they can develop monitoring and troubleshooting systems more rapidly in such scenarios.

11.1.2 Brownfield approach

This approach involves a gradual transition from 4G to 5G, where the existing 4G infrastructure remains operational without immediate disruption. Instead, 5G radio and core network nodes are introduced as independent units alongside the 4G infrastructure, enabling the provision of 5G coverage and services in specific regions. This enables the operator to swiftly deliver 5G services to end users, primarily catering to enhanced mobile broadband (eMBB) use cases. To support more complex scenarios like network slicing and MEC use cases, planning and implementation of distributed data centers, enhancements to the IP core, automation capabilities, and virtualized radio network readiness become essential.

Figure 11.1: SA and NSA deployment options.

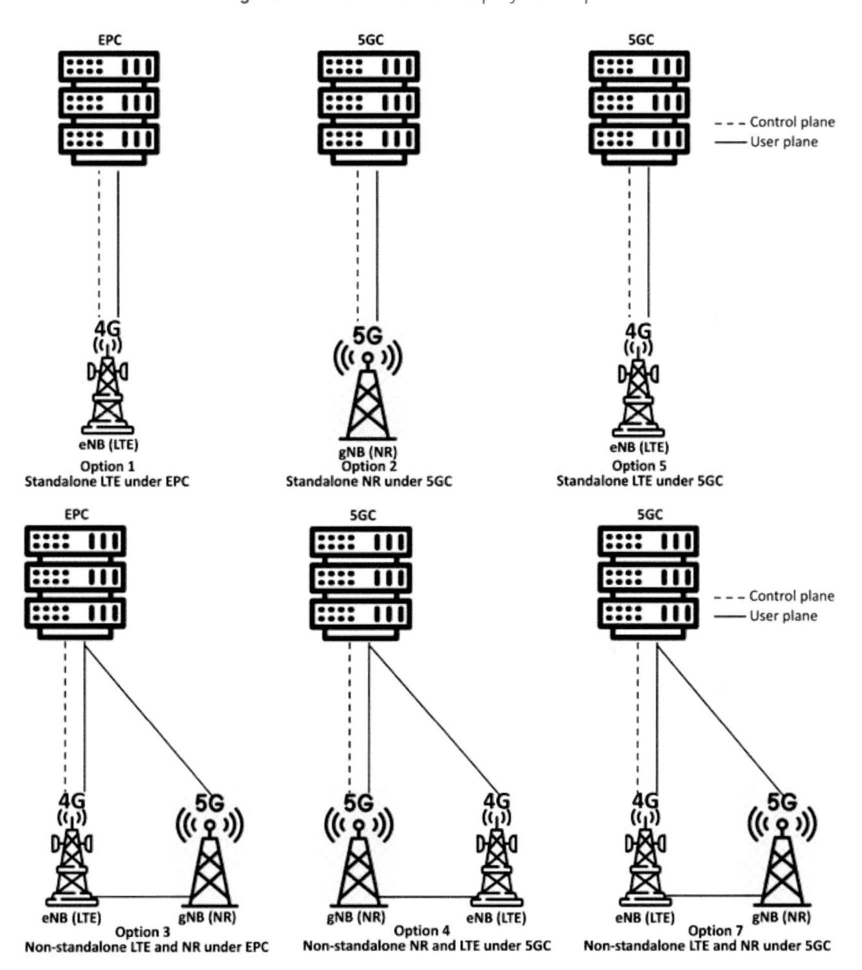

11.2 Migration Path from 4G to 5G

The migration path selected by majority of the operators (with existing 4G network) is to go from Option 1 to Option 3x and finally to Option 2. Option 3x refers to Option 3 with the data split happening at the en-gNB. The data traffic passes directly between the en-gNB and the SGW with option to share the load with eNB if required.

11.2.1 Option 1 to Option 3x (Figure 11.2)

Option 1 is the existing 4G deployment, whereas NSA Option 3x is based on the EN-DC, which requires NR en-gNB in E-UTRAN and newer features on LTE eNB to support EN-DC procedures. Option 3x allows compatible devices to use dual connectivity to combine LTE and NR radio access. With this the excessive user plane traffic on eNB is eliminated. The eNB acts as the master node and

Figure 11.2: Option 1 to Option 3x.

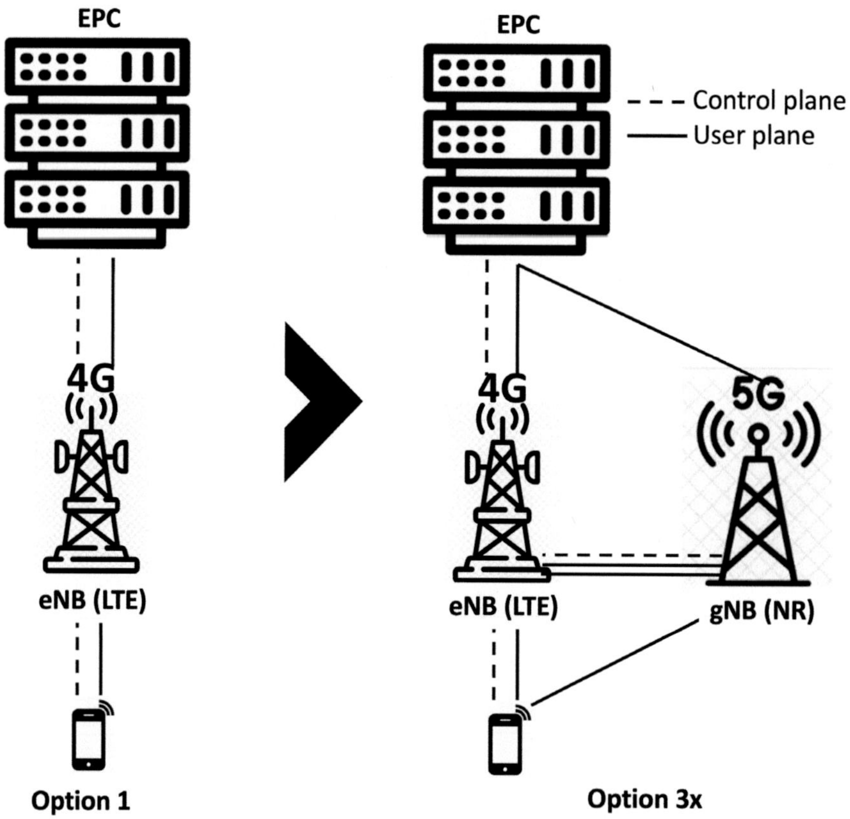

controls the traffic split between en-gNB and eNB allowing a seamless mobility and service continuity between 4G and 5G.

This deployment option will be very quick to go to market and helps in adding capacity in areas where traditional 4G coverage is missing. Operators can add small cell deployments in the 5G high-band spectrum to achieve this.

Per subscriber throughput will increase with NR. So the network needs to be planned to accept higher throughputs (per subscriber and in total). Operators should consider adding more user-planes into their network and even go for options like MEC to break out the traffic to avoid congesting their IP transport. The existing IP transport network itself might need upgrades to support higher bandwidth and throughput.

Since the UE will continue to use the 4G/EPC core, no additional support is required on the devices except the support for using NR as non-stand alone access together with LTE.

4G core capacity needs to be enhanced by adopting the control and user plane separation (CUPS). Additionally, the core also needs to support signaling on different interfaces for the higher 5G bandwidth and QoS. New attributes on policy interfaces for the 5G throughput and QoS must also be supported by the core.

In Option 3x, network decides which bearers should use EN-DC. Voice services (VOLTE) don't require high bandwidth and can remain on the 4G eNB initially.

The technical details of Option 3x NSA are provided in Chapter 7.

11.2.2 Option 3x to Option 2 + Option3x (Figure 11.3)

Operators will introduce 5G to the customers in the form of NSA. It is expected that the entire network will get converted to NSA gradually, before the SA adoption starts. In this section we look at how an NSA network gets converted to a pure 5G SA network. Note that it is not a single step process and will involve NSA and SA existing together for some time. This will require the devices to support inter-RAT mobility between LTE/NR (NSA) to NR only (SA).

Adopting Option 2 allows to use the full potential of 5G core and supports the new use-cases defined for 5G. Operators could even consider moving the services on existing EPC to 5GC.

Figure 11.3: Option 3x to Option 3x + Option 2.

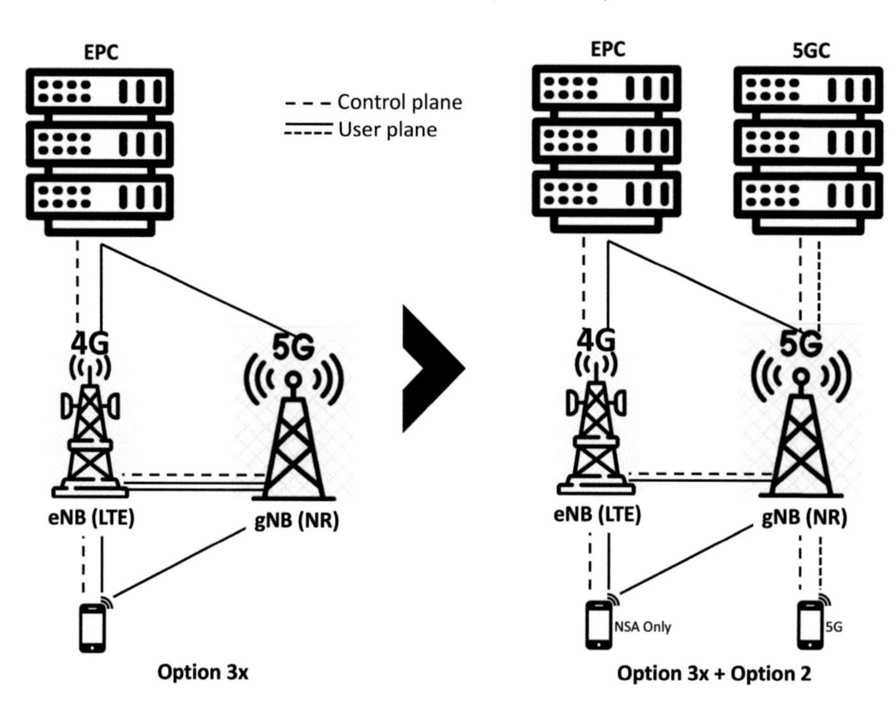

Operators should consider moving to Option 2 once there is a wide area coverage for the NR and can assure seamless handover between 5GC and EPC.

Option 2 requires update of NR gNB to support both NSA (Option 3x) and SA (Option 2) in parallel. If the NR gNBs can connect to both 5GC and EPC during the transition phase, 5G devices supporting only NSA can still be used with their 5G radio capabilities. Option 2 also requires the eNodeBs to be connected to EPC to support inter-RAT mobility. On the NR/eUTRAN side, proper planning will be required to assure that the gNBs connected to the 4G core as well as 5G core prioritize the procedures like addition of gNB from NSA or support intra-5G handovers.

From a core perspective, N26 (interface between AMF and MME) will be required for seamless mobility. N26 can be used to transfer the subscriber's authentication and session information between the EPC and 5GC.

From a session continuity perspective, IP address retention during mobility is required. This calls for converged SPGW-C/SMF and SPGW-U/UPF nodes.

Operators supporting VoLTE can either remain to provide voice services on 4G or move to use VoNR using the 5GC. However, this complete shift will require extensive testing of voice services in 5G. For customers using 4G devices, VoLTE still needs to be supported.

11.3 Migration Considerations

With Option 3x and Option 2 running in parallel to cater to different types of devices, there are certain considerations that should be handled by the operator to make sure that the right device goes to the right core. Having both options together requires some changes in the MME/DNS to select the right node from 4G (SAEGW) or 5G (converged PGW+SMF) for 5G SA capable UE trying to establish a connection in 4G with an appropriate APN. The network/node selection algorithm should consider the following parameters to ensure seamless interworking in case the UE moves from 4G to 5G coverage area and vice versa.

- UE capability (4G / 5GNSA / 5GSA)
- UE subscription
- UE selected APN/DNN.

Operators can plan a separate DNN for 5G which can then be used as an input for the core selection. In addition, based on capacity and load on the EPC side, the decision can be taken whether to send the subscribers to EPC or to the converged core for the device attaching on 4G.

Ideally only a subscriber with a 5G subscription asking for a 5G APN/DNN should be allowed to attach to the converged PGW+SMF node. With slicing in 5G, it is important that the MME chooses the right network taking into consideration the UE's slice subscription also. Choosing a converged PGW+SMF/PGW during the initial attach and PDU session activation (in 4G) which does not support the subscribed slice will cause service disruption during inter-RAT mobility to 5G.

Glossary

The following abbreviations are used in this document.

Term	Description
5GC	5G Core network
AF	Application function
AI	Artificial intelligence
AMF	Access and Mobility Management Function
API	Application programmable interface
AUSF	Authentication server function
BBU	Baseband unit
BGP	Border gateway protocol
BP	Branching point
CBC	Cell broadcasting center
CG-NAT	Carrier grade – network address translation
CHF	Charging function
CLI	Command line interface
CNF	Containerized network function
CPF	Control plane function
CPRI	Common public radio interface
CRAN	Cloud/centralized RAN
CU	Centralised Unit
CUPS	Control and User Plane Separation
DC	Data center
DNAI	Data network access identifier
DNN	Data network name
DNS	Domain name system
DPI	Deep packet inspection
DRA	Diameter routing agent
DU	Distributed unit
eCPRI	Enhanced common public radio interface
eMBB	Enhanced mobile broadband

eNB	Evolved node B
EAS	Edge application server
EPC	Evolved packet core
EPRTC	Enhanced primary reference time clock
EVPN	Ethernet VPN
FQDN	Fully qualified domain name
gNB	Next generation node B
IGP	Interior gateway protocol
IMSI	International mobile subscriber identity
IoT	Internet of Things
KPI	Key performance indicator
LADN	Local area data network
L3VPN	Layer 3 virtual private network
LCM	Life-cycle Management/Manager
LTE	Long term evolution
MANO	Management and organization
MEC	Multi-access edge computing
mMTC	Massive machine type communications
MNO	Mobile network operator
MPLS	Multi-protocol label switching
MVNO	Mobile virtual network operator
NEF	Network exposure function
NF	Network function
NFV	Network function virtualization
NFVI	NFV infrastructure
NFVO	NFV orchestration
NR	New radio
NRF	NF repository function
NSA	Non-standalone (5G)
NSD	Network service descriptor
NSI	Network slice instance
NSMF	Network slice management function
NSSF	Network slice selection function
NSO	Network service orchestration
OAM	Operation, Administration and Management
OEC	Optical to Electrical Converter
PCC	Policy and Charging Control
PCF	Policy control function
PDN	Packet data network
PDU	Protocol data unit
PEI	Permanent equipment identifier
PGW	Packet gateway

PRC	Primary reference clock
PRTC	Primary reference time clock
PTP	Precision time protocol
QoS	Quality of service
RAN	Radio access network
RLC	Radio link control
RRH	Remote radio head
RU	Remote unit
SA	Standalone (5G)
SBA	Service based architecture
SBI	Service based interface
SDN	Software-defined network
SLA	Service-level agreement
SMF	Session management function
S-NSSAI	Single network slice selection assistance information
SP	Service provider
SR	Segment routing
SR-PCE	Segment routing – path computation element
SSC	Session and service continuity
SUPI	Subscription permanent identifier
TBC	Telecom boundary clock
TE	Traffic engineering
TN	Transport network
ToR	Top of rack
UDM	Unified data management
UDR	Unified data repository
UE	User equipment
UL-CL	Uplink classifier function
UPF	User plane function
URLLC	Ultra reliable low latency communication
URSP	UE route selection policy
V2X	Vehicle to everything
VIM	Virtualization infrastructure manager
VNF	Virtual network function
VNFM	VNF manager
VPN	Virtual private network
VRF	Virtual routing function
xHaul	Any name + haul (front-haul, mid-haul, back-haul, etc.)

References

[1] Pros and cons of various cluster approaches https://learnk8s.io/how-many-clusters https://www.3gpp.org/dynareport?code=38-series.htm https://www.5gtechnologyworld.com/how-timing-propagates-in-a-5g-network/

[2] Ref: O-RAN.WG4.CUS.0-v05.00. O-RAN Fronthaul Working Group

[3] CPRI splits with latency and bandwidth-https://www.3gpp.org/news-events/3gpp-news/open-ran

[4] Ref: O-RAN, 2021. Control, User and Synchronization Plane Specification: O-RAN.WG4.CUS.0-v05.00. O-RAN Fronthaul Working Group.

[5] Ref: O-RAN, 2021. Control, User and Synchronization Plane Specification: O-RAN.WG4.CUS.0-v05.00. O-RAN Fronthaul Working Group

[6] Telco Cloud: Why it hasn't delivered, and what must change for 5g. STL Partners

[7] The Journey to Telco Cloud - What to Consider When Evolving Networks to the Cloud. Openet Amdocs

[8] Exploring the Benefits of Disaggregated, Cloud-Native Architectures for Telco Cloud Transformation. IDC Analyst Connection

[9] 3GPP,"System architecture for the 5G System(5GS),"TS23.501Version 16.9.0, 3rd Generation Partnership Project (3GPP), June 2021.

[10] 3GPP,"Procedures for the 5G System(5GS),"TS23.502 Version 16.9.0, 3rd Generation Partnership Project (3GPP), June 2021.

[11] Anil Rao. 5G network slicing: cross-domain orchestration and management will drive commercialization. *Analysys Mason*

[12] *O-RAN.WG1.Slicing-Architecture-R003-v11.00*

[13] Salah Eddine Elayoubi, Sana Benjemaa, Zwi Altman, Ana Maria Galindo Serrano. 5G RAN slicing for verticals: Enablers and challenges. IEEE Communications Magazine, 2019, 57 (1), pp.28-34.

[14] ONAP Network Slicing (and related) Features in 3GPP

[15] Anand R. Prasad, Sivabalan Arumugam, Sheeba B and Alf Zugenmaier, “3GPP 5G Security”, Journal of ICT Standardization (River Publishers, Vol. 6, Iss. 1&2).

[16] Road to 5G: Introduction and Migration. GSMA

Index

U

V

X

Printed in the United States
by Baker & Taylor Publisher Services